Examination of the U.S. Air Force's **Aircraft Sustainment Needs in the Future** and Its Strategy to Meet Those Needs

Committee on Examination of the U.S. Air Force's Aircraft Sustainment Needs in the Future and Its Strategy to Meet Those Needs

Air Force Studies Board

Division on Engineering and Physical Sciences

NATIONAL RESEARCH COUNCIL
OF THE NATIONAL ACADEMIES

THE NATIONAL ACADEMIES PRESS
Washington, D.C.
www.nap.edu

THE NATIONAL ACADEMIES PRESS **500 Fifth Street, N.W.** **Washington, DC 20001**

NOTICE: The project that is the subject of this report was approved by the Governing Board of the National Research Council, whose members are drawn from the councils of the National Academy of Sciences, the National Academy of Engineering, and the Institute of Medicine. The members of the committee responsible for the report were chosen for their special competences and with regard for appropriate balance.

This is a report of work supported by Grant FA9550-10-1-0411 between the U.S. Air Force and the National Academy of Sciences. Any opinions, findings, conclusions, or recommendations expressed in this publication are those of the author(s) and do not necessarily reflect the view of the organizations or agencies that provided support for the project.

International Standard Book Number-13: 978-0-309-21520-6
International Standard Book Number-10: 0-309-21520-X

Limited copies of this report are available from:

Air Force Studies Board
National Research Council
500 Fifth Street, N.W.
Washington, DC 20001
(202) 334-3111

Additional copies are available from:

The National Academies Press
500 Fifth Street, N.W.
Lockbox 285
Washington, DC 20055
(800) 624-6242 or (202) 334-3313
(in the Washington metropolitan area)
Internet, http://www.nap.edu

Printed in the United States of America

COMMITTEE ON EXAMINATION OF THE U.S. AIR FORCE'S AIRCRAFT SUSTAINMENT NEEDS IN THE FUTURE AND ITS STRATEGY TO MEET THOSE NEEDS

S. MICHAEL HUDSON, Rolls-Royce North America (retired), *Co-Chair*
MICHAEL E. ZETTLER, Z-Zettler Consulting, *Co-Chair*
MEYER J. BENZAKEIN, The Ohio State University
CHARLES E. BROWNING, University of Dayton
DIANNE CHONG, The Boeing Company
DAVID E. CROW, University of Connecticut
FRANK R. FAYKES, U.S. Air Force (retired)
JOHN T. FOREMAN, Software Engineering Institute, Carnegie Mellon University
WESLEY L. HARRIS, Massachusetts Institute of Technology
HOWARD F. HETRICK, Northrop Grumman Corporation
CLYDE KIZER, Airbus North America (retired)
THOMAS A. McDERMOTT, JR., Georgia Tech Research Institute
LYLE H. SCHWARTZ, University of Maryland
BRUCE M. THOMPSON, Sandia National Laboratories
RAYMOND VALEIKA, Delta Airlines (retired)

Staff

CARTER W. FORD, Study Director
KAMARA E. BROWN, Research Associate
SARAH M. CAPOTE, Research Associate
NORMAN M. HALLER, Consultant
ZEIDA PATMON, Program Associate
MARGUERITE E. SCHNEIDER, Administrative Coordinator

Preface

Having now been at war for two decades, the U.S. Air Force is finding that its legacy aircraft are becoming increasingly more expensive to operate and maintain. Looking ahead, and facing a constrained overall budget, the Air Force is concerned that the resources needed to sustain its legacy aircraft may increase to the point where they could consume the resources needed to modernize the Air Force. Recognizing the importance of sustainment, both to the accomplishment of its current wartime missions and to the potential capabilities of its future aircraft, the Air Force asked the National Research Council (NRC) of the National Academies to conduct this study of sustainment.

A committee of experts with significant experience in both technical and operational areas related to sustainment was formed to conduct this study. Meeting for the first time in October 2010, the committee quickly grasped the complexities inherent in the terms of reference, which addressed a substantial portion of overall Air Force activities and resources—both current and future. Because of the need for solid data to supplement its own knowledge and capabilities, the committee is very grateful for the responsive and highly informed cooperation of numerous representatives from the Air Force as well as from government, industry, and academia. The committee co-chairs especially wish to thank all of the committee members for their many insightful contributions and tireless efforts in producing this report on schedule.

S. Michael Hudson, *Co-Chair*
Michael E. Zetter, *Co-Chair*
Committee on Examination of the U.S. Air Force's Aircraft Sustainment Needs in the Future and Its Strategy to Meet Those Needs

Acknowledgment of Reviewers

This report has been reviewed in draft form by individuals chosen for their diverse perspectives and technical expertise, in accordance with procedures approved by the National Research Council's Report Review Committee. The purpose of this independent review is to provide candid and critical comments that will assist the institution in making its published report as sound as possible and to ensure that the report meets institutional standards for objectivity, evidence, and responsiveness to the study charge. The review comments and draft manuscript remain confidential to protect the integrity of the deliberative process. We wish to thank the following individuals for their review of this report:

John-Paul B. Clarke, Georgia Institute of Technology,
Stephen P. Condon, Maj Gen, USAF (retired), Dayton Aerospace, Inc.,
Lawrence J. Delaney, Titan Corporation (retired),
Elisabeth M. Drake, Massachusetts Institute of Technology,
Lester L. Lyles, Gen, USAF (retired), Independent Aerospace Consultant,
Lisa Mahlmann, Lockheed Martin Aeronautics Company,
David Miller, Massachusetts Institute of Technology,
Charles F. Tiffany, The Boeing Company (retired), and
Stephen Wei-Lun Tsai, Stanford University.

Although the reviewers listed above provided many constructive comments and suggestions, they were not asked to endorse the conclusions or recommendations, nor did they see the final draft of the report before its release. The review of this report was overseen by Sheila E. Widnall, Massachusetts Institute of Technology. Appointed by the National Research Council, she was responsible for making certain that an independent examination of this report was carried out in accordance with institutional procedures and that all review comments were carefully considered. Responsibility for the final content of this report rests entirely with the authoring committee and the institution.

Contents

SUMMARY 1

1 INTRODUCTION 15
Overview, 15
Committee Formation and Terms of Reference, 16
Study Approach, 16
Defining Sustainment, 18
Weapon System Sustainment in the Context of the Military Mission, 18
Weapon System Sustainment Defined for This Study, 19
Past, Present, and Future Environments for Weapon System Sustainment, 19
Military Operations, 19
Weapon System Sustainment Goals of the Air Force, 20
Governance: Laws, Policies, Strategies, and Regulations, 20
Relationships, 21
Budget, 21
Report Organization, 22

2 REVIEW OF THE IMPACT OF REGULATIONS, POLICIES, AND STRATEGIES ON SUSTAINMENT 24
Introduction, 24
Air Force Sustainment Goals, 25
Statutes, Regulations, and Policies That Impact Air Force Sustainment, 26

Overview, 26
Statutes and Regulations, 28
Establishing Air Force Sustainment Policies, 33
Engineering-based Decisions, 34
Comprehensive Approach to Gathering and Analyzing Data, 36
Current Governance Structure for Air Force Sustainment, 37
Current Air Force Organization Complexities for Collaboration, 44
Outcome-based Metrics, 45
Past, Present, and Potential Future Strategies for Air Force Sustainment, 48
A Benchmark: Processes and Policies of the Naval Aviation Enterprise, 57
Concluding Thoughts, 63

3 ASSESSMENT OF CURRENT SUSTAINMENT INVESTMENTS, INFRASTRUCTURE, AND PROCESSES 65
Introduction, 65
Air Force Investment Process Relating to Sustainment, 65
Current Resources and Investments, 70
Air Force Sustainment Infrastructure, 74
Investment in Infrastructure, 75
Ground Equipment, 77
Current Air Force Sustainment Processes, 77
Workforce, 78
Acquisition, 79
Supply Chain, 80
Maintenance Processes, 81
Resourcing Processes, 82
Fleet Viability Board, 83
Logistics Support, 84
eLog21, 84
Obsolescence and Diminishing Manufacturing Sources, 85
Concluding Thoughts, 86

4 ASSESSMENT OF AIR FORCE AIR LOGISTICS CENTERS 87
Introduction, 87
Methodology for the Assessment, 91
History and Present State of Air Force Air Logistics Centers, 92
Holistic Approach to the Responsibilities and Performance of Air Force Air Logistics Centers, 97
Assessment of the Resourcing of Air Logistics Centers, 99
Policy-Driven Supply Support, 99
Resourcing the ALC Workforce, 100

50/50 Affects the ALC Workforce, 101
Critical Support Staff for the ALC Workforce,102
Workforce Training, 103
Labor Relations and Growing the Workforce, 103
Organizing the Air Logistics Centers for Sustainment, 104
Resourcing for Technology Insertion at Air Logistics Centers, 113
Enterprise Management at the Air Logistics Centers, 114
Maintenance Planning and Priorities, 116
Software Sustainment for Legacy and Future Systems, 117
Current Trends in Air Force Software Development and Maintenance, 118
Air Force Policies for Software Sustainment, 119
Software Workforce, 121
Software Facilities, 122
Organization and Management, 123
Future Challenges Based on New Aircraft Entering the Inventory, 123
Sustainment Funding, 126
Efficiency and Effectiveness of Air Force Air Logistics Centers, 127
Concluding Thoughts, 128

5 TECHNOLOGY DEVELOPMENT AND INSERTION FOR SUSTAINMENT 132
Introduction, 132
Policies and Guidance, 134
Technology Development and Transition, 136
Technology Areas Relevant to Sustainment, 138
Defining Sustainment Technology Needs, 138
Long-Term Research, 143
Air Force Sustainment Technology, 146
Overview, 146
Historical Background, 147
Changing Strategies and Processes, 150
Summary of Recent History, 157
Transition to the Future, 158
Requirements, 159
Resources, 164
Right People, 167
Recommendation, 168
Concluding Thoughts, 168

6 INCORPORATING SUSTAINABILITY INTO FUTURE DESIGNS 170
Introduction, 170
Incorporating Sustainability in the Concept and Initial Planning, System Design and Development, and Deployment and Support Phases, 171
Incorporating Desirable Design Features and Applying Lessons Learned During Weapon System Life-Cycle Phases, 176
Data Rights/Access and the Air Force's Ability to Gain Weapon System Sustainment Domain Knowledge, 179
Considering a Blended Support Concept, 180
Moving to a Data-Driven Sustainment Strategy and Common Enterprise Management in New Designs, 181
The Data or Information Systems Needed to Develop the Maintenance Program, 182
Information and Data Needs Once a Platform Enters Service, 182
Information System Needed for Continuing Analysis and Surveillance of the Sustainment Process, 183
Providing for Continued Incorporation of Technology for Sustainment, 185
Overview, 185
Software Systems, 185
Air Vehicles and Engines, 185
Integrity Programs, 186
The Unique Sustainment Aspects with Respect to Rapidly Fielded Systems, 186
Commercial Aviation Practices for Air Force Consideration, 187
Delegation, 188
Aging Aircraft Sustainment, 189
Maintenance Program Development, 189
Concluding Thoughts, 190

APPENDIXES

A Biographical Sketches of Committee Members 195
B Meetings and Participating Organizations 204
C Navy Enterprise Transformation: Working for the Greater Good 215

Acronyms

AA	aircraft availability
ACC	Air Combat Command
AD	Airworthiness Directive
AETC	Air Education and Training Command
AF	Air Force
AFGLSC	Air Force Global Logistics Support Center
AFI	Air Force Instruction
AFMC	Air Force Materiel Command
AFOSR	Air Force Office of Scientific Research
AFPD	Air Force Policy Directive
AFRL	Air Force Research Laboratory
AFROC	Air Force Requirements Oversight Council
ALC	Air Logistics Center
AMC	Air Mobility Command
AMC/A4	Air Mobility Command Directorate of Logistics
AMOC	Alternative Means of Compliance
ASIP	Aircraft Structural Integrity Program
BCA	Business Case Analysis
BOD	Board of Directors
BRAC	Base Realignment and Closure

CAMS	Core Automated Maintenance System
CASS	Continuing Analysis and Surveillance System
CBM	Condition-Based Maintenance
CFFC	Commander, U.S. Fleet Forces Command
CLS	Contractor Logistics Support
CMMI	Capability Maturity Model Integration
CNAF	Commander, Naval Air Forces
CNO	Chief of Naval Operations
DAMS	Defense Acquisition Management System
DAR	Designated Airworthiness Representative
DAU	Defense Acquisition University
DER	Designated Engineering Representative
DMS/MS	Diminishing Manufacturing Sources and Materiel Shortages
DoD	Department of Defense
eLog21	Expeditionary Logistics for the Twenty-first Century
EO	Engineering Order
ERP	Enterprise Resource Planning
FAA	Federal Aviation Administration
FAR	Federal Air Regulation
FBO	Fixed Base Operator
FMECA	Failure Mode, Effect and Criticality Assessment
FYDP	Future Years Defense Program
GM	General Motors
GSP	Globemaster Sustainment Partnership
ILCM	Integrated Life-Cycle Management
ISG	Industry Steering Groups
LMI	Logistics Management Institute
LogEA	(Air Force) Logistics Enterprise Architecture
LRT	Logistics Requirements Traceability
MAJCOM	Major Command
ManTech	manufacturing technology
MICAP	Mission Incapable Awaiting Parts
MIL-STD	Military Standard
MMEL	Master Minimum Equipment Lists

MRB	Maintenance Review Board
MRO	Maintenance, Repair and Overhaul
MSG-3	Maintenance Steering Group Three
MXW	Maintenance Wing
NAE	Naval Aviation Enterprise
NAVAIR	Naval Air Systems Command
NDI	Non-destructive Inspection
NRC	National Research Council
O&M	operations and maintenance
O&S	operations and support
OC-ALC	Oklahoma City Air Logistics Center
OEM	original equipment manufacturer
ONR	Office of Naval Research
OO-ALC	Ogden Air Logistics Center
OPNAV	Office of the Chief of Naval Operations
OSD	Office of the Secretary of Defense
PEO	Program Executive Office
POM	Program Objective Memorandum
PBL	Performance-based Logistics
RCM	Reliability Centered Maintenance
RDT&E	research, development, testing, and engineering
REMIS	Reliability and Maintainability Information Systems
S&T	science and technology
SAE	Service Acquisition Executive
SAF/AQ	Assistant Secretary of the Air Force (Acquisition)
SAF/AQX	Deputy Assistant Secretary of the Air Force (Acquisition Integration)
SAF/IE	Assistant Secretary of the Air Force (Installations, Environment and Logistics)
SECDEF	Secretary of Defense
SFDM	Single Fleet, Driven Metric
SIL	software integration laboratory
SLOC	Source Lines of Code
SPM	System Program Manager
SPO	System Program Office; Single Process Owner

TOA	Table of Allowance
TOR	terms of reference
TSPR	Total System Performance Requirement
TSSR	Total System Support Requirement
USAF	U.S. Air Force
WR-ALC	Warner-Robins Air Logistics Center

Summary

The U.S. Air Force weapon system sustainment[1] enterprise is extremely large in terms of scope, workforce, and associated costs that amount to billions of dollars annually. When other expenditures, such as selected personnel accounts, fuel, facilities, and utilities are included, the total dollar amount expended is significantly higher.[2] As a point of reference for the reader, the Air Force total costs of sustainment activities exceed the operating costs of such industry giants as American Airlines and Delta Airlines.[3] As demonstrated by the following comment by the Under Secretary of Defense for Acquisition, Logistics, and Technology, sustainment is a major component of the Department of Defense (DoD) budget:

> About $100 billion is procurement of weapons systems—OK, acquisition—but *remember that 70 percent of the cost of a weapons system is not acquiring it; it is sustaining it.* It's not buying it; it's having it, where most of the money is. Said differently, *most of the money in*

[1]Adopted from the 2009 report *DoD Weapon System Acquisition Reform Product Support Assessment.* This report applies the following definition of weapon system or aircraft sustainment: System sustainment is the package of support functions required to maintain the readiness and operational capability of weapon systems, subsystems, software, and support systems. It encompasses materiel management, distribution, technical data management, maintenance, training, cataloging, configuration management, engineering support, repair parts management, failure reporting and analysis, and reliability growth. Available at https://acc.dau.mil/adl/en-US/328610/file/47489/DoD%20Weapon%20System%20Acquisition%20Reform%20PSA_19%20NOV_Final.pdf. Accessed August 18, 2011.

[2]Chapter 3 includes an in-depth discussion of the Air Force expenditures.

[3]Chapter 6 includes a discussion of commercial best practices for sustainment.

> *the budget is spent on sustaining weapons systems that were procured in the past rather than on acquisition programs per se.* And we can't leave that much money out of the better buying power equation [emphasis added].[4]

In May 2010, DoD introduced the Defense Efficiencies Initiative that seeks to "increase efficiencies, reduce overhead costs, and eliminate redundant functions in order to improve the effectiveness of the DoD enterprise. This effort is focused on reprioritizing how DoD can use resources to more effectively support and *sustain* the force [emphasis added]."[5] The Defense Efficiencies Initiative recognizes the need for improved efficiency and effectiveness to support and sustain the force. This study highlights many sustainment issues and offers recommendations aimed at improving the efficiency and effectiveness of the Air Force weapon system sustainment enterprise.

The Air Force weapon system sustainment enterprise consists of a highly skilled workforce but operates without modern enterprise resource planning tools and with a supply chain that is not structured according to business best practices. In fact, Air Force weapon system sustainment, including the demand to maintain aircraft organically and by contractor logistics support, dwarfs that of commercial airlines and other nations' Air Forces. The sustainment posture is determined by the number and variety of aircraft, the technology of the systems involved, and the global deployment of the fleet. The fleet's diversity, which ranges from aircraft designed and deployed in the 1950s to the world's most advanced high-performance fighters, weighs on the enterprise's operation. The enterprise has become more complex over time not only because of the fleet's increased growth and diversity, but also because of global politics and regulations.

The Air Force has been operating on a wartime-like footing for the past 20 years. This extended period of intense operation has been further complicated by multiple theaters of operation and by demand for more diverse and long-duration fighter missions; worldwide airlift; long-range bomber sorties; diverse intelligence, surveillance, and reconnaissance aircraft; and extensive aerial refueling sorties resulting from increasing geopolitical complexity.

Today, sustainment activities are undertaken by numerous offices and organizations, including the Air Force Secretariat; the Air Force Air Staff; the Air Force Materiel Command (AFMC) and its subordinate Product Centers; Air Logistics

[4]Ashton Carter, Under Secretary of Defense for Acquisition, Technology, and Logistics. 2011. "Pentagon Efficiency Initiatives." Remarks given at the Heritage Foundation, Washington, D.C., April 20. Available at http://www.heritage.org/Events/2011/04/Pentagon-Efficiency?query=Pentagon+Efficiency+Initiatives:+Are+They+Enough+to+Stave+Off+More+Defense+Cuts?. Accessed May 2, 2011.

[5]DoD. 2010. Defense Efficiencies Initiative. Available at http://www.defense.gov/home/features/2010/0810_effinit/. Accessed April 21, 2011.

Centers (ALCs); and Air Force operational commands. The activities of all of these offices and organizations must be considered in studies of weapon system sustainment. Sustainment is generally program-specific by weapon system. Although policy is made by the Secretariat and Air Staff offices, it is broad and open to interpretation. Sustainment activities may occur in the operating command, but they are generally short-term in nature. The Product Centers and ALCs drive the larger-scale and high-cost sustainment activities. Because of the broad nature of sustainment policy and the number of organizations involved there is no "model" weapon system sustainment program.

Air Force weapon system sustainment has functioned under a variety of concepts and organizations, yet the system has repeatedly met national and global threats, largely because of the dedication of the men and women responsible for the detailed tasks of sustainment. Sustainment activities require significant coordination and communication across a myriad of functions and organizations. At present, this process is largely facilitated by interpersonal relationships rather than clear, concise lines of authority and modern enterprise reporting and planning tools, which results in escalating costs and inefficiencies.

The Air Force's sustainment activities achieve the desired operational outcomes. However, extraordinary management effort and attention are required to knock down stovepipes that impede efficiencies. Senior officials are consistently frustrated by weak or overly broad policies, minimal governance, and unnecessarily complex organizational structures as they try to improve support postures resulting from enterprise inefficiencies. These systemic shortcomings span the weapon system life cycle—from initial concept development through retirement—for which there is a lack of clear accountability. For the foreseeable future, the Air Force will continue to operate at a high tempo but will face tremendous pressures on its sustainment budget. The increasing costs of sustainment are a concern, because, as the present systems continue to age, they will demand a greater proportion of the overall defense budget. This may not be "sustainable" in the context of the downward pressures on the overall defense budget, shown in Figure S-1.[6]

STUDY APPROACH

In response to a request from the Deputy Assistant Secretary of the Air Force for Science, Technology, and Engineering, the National Research Council, under the auspices of the Air Force Studies Board, formed the Committee on Examination

[6]Blaise J. Durante, Deputy Assistant Secretary for Acquisition Integration, Office of the Assistant Secretary of the Air Force for Acquisition. "Budgeting Considerations Related to Sustainment." Presentation to the committee, October 21, 2010.

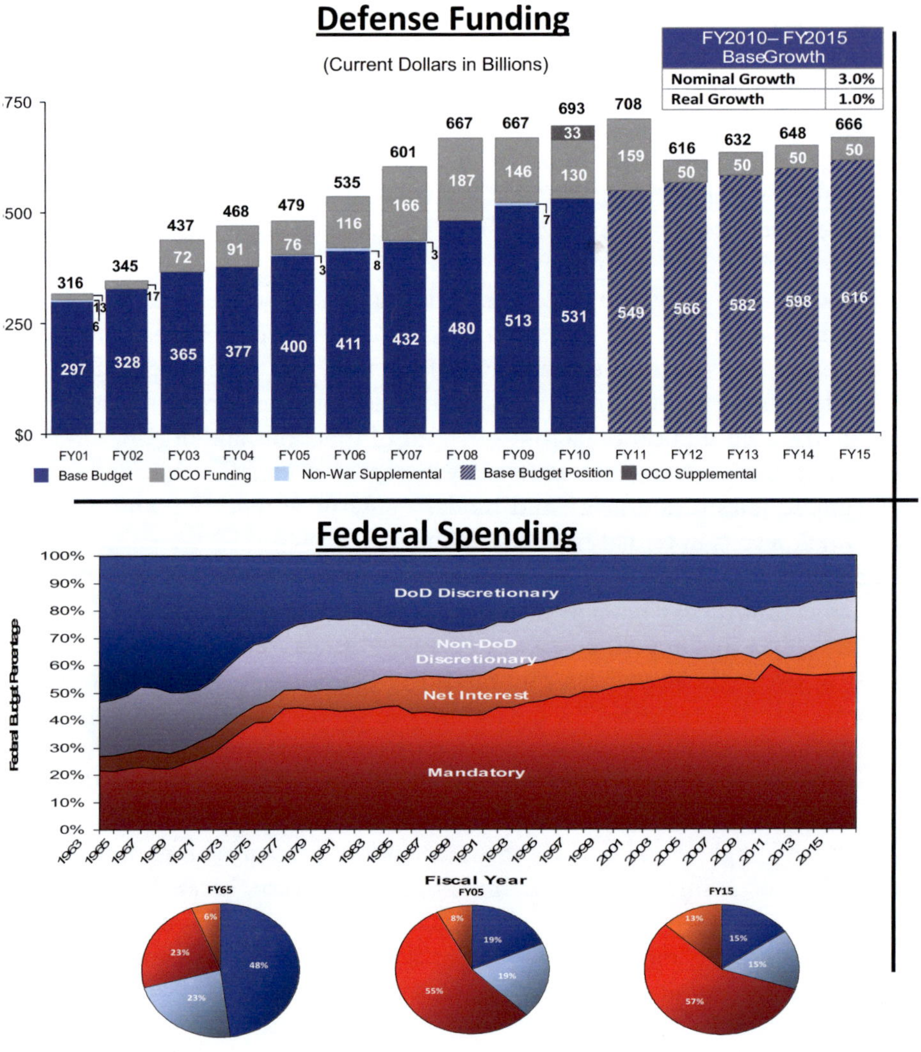

FIGURE S-1
Economic realities will affect defense spending. OCO, Overseas Contingency Operations. SOURCE: Blaise J. Durante, Deputy Assistant Secretary for Acquisition Integration, Office of the Assistant Secretary of the Air Force for Acquisition. "Budgeting Considerations Related to Sustainment." Presentation to the committee, October 21, 2010.

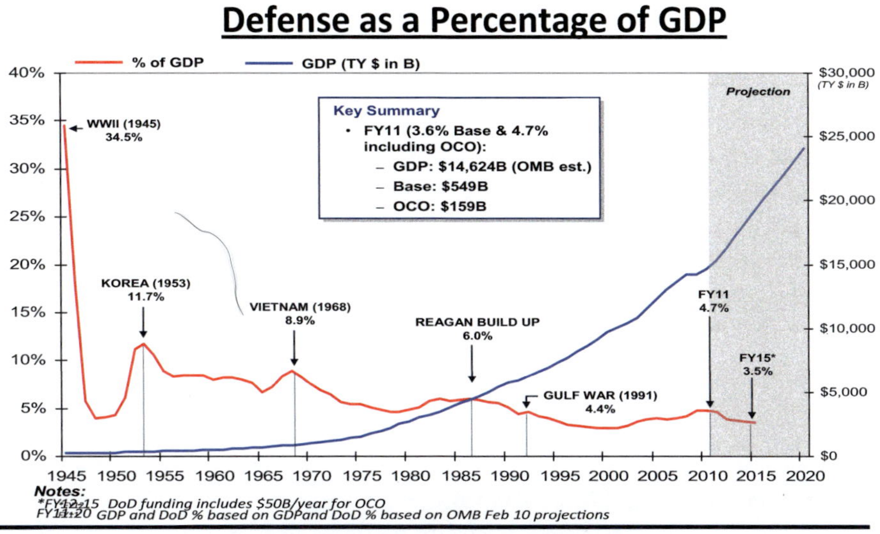

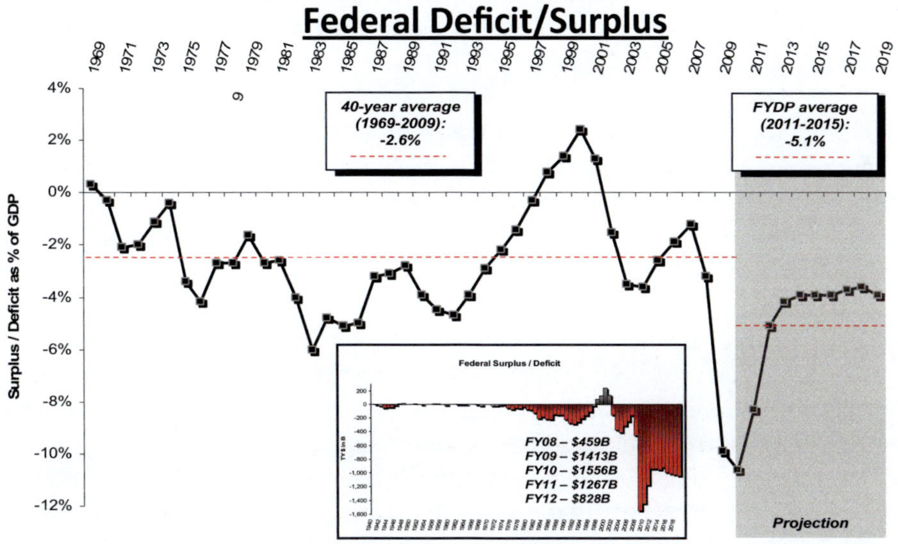

FIGURE S-1 *continued*

of the U.S. Air Force's Aircraft Sustainment Needs in the Future and Its Strategy to Meet Those Needs to address the following terms of reference:[7]

- Assess current sustainment investments, infrastructure, and processes for adequacy in sustaining aging legacy systems and their support equipment.
- Determine if any modifications in policy are required and, if so, identify them and make recommendations for changes in Air Force regulations, policies, and strategies to accomplish the sustainment goals of the Air Force.
- Determine if any modifications in technology efforts are required and, if so, identify them and make recommendations regarding the technology efforts that should be pursued because they could make positive impacts on the sustainment of the current and future systems and equipment of the Air Force.
- Determine if the Air Logistics Centers have the necessary resources (funding, manpower, skill sets, and technologies) and are equipped and organized to sustain legacy systems and equipment and the Air Force of tomorrow.
- Identify and make recommendations regarding incorporating sustainability into future aircraft designs.

During four data-gathering meetings, senior Air Force leaders, including representatives of several Air Force Major Commands, representatives from the other military departments, senior officials in the Office of the Secretary of Defense, professional staff members from key congressional oversight committees, and senior industry executives provided input to the committee (see Appendix A for biographical sketches of the committee members and Appendix B for meetings and participating organizations). Additionally, the committee held smaller site visits with Air Force and Navy officials, including visits to the three ALCs and the Fleet Readiness Center Southwest. Finally, the committee held two 3-day meetings to finalize its report and findings and recommendations.

The following "Big 7" themes continuously recurred throughout the study and correspond to the recommendations cited: (1) the lack of Air Force processes for sustainment as an enterprise (Recommendations 2-2, 2-3, 3-1, 4-5, 4-7, 4-8, 6-1, 6-3, 6-5, and 6-10); (2) ill-defined Air Force organizational structure for sustain-

[7]Many of the presenters to the committee agreed that a budget "train wreck" with respect to sustainment costs is looming. The committee engaged in considerable discussion on how to validate the spontaneous comments regarding out-of-control sustainment costs. The terms of reference did not require the committee to undertake business case analyses related to the report recommendations. Likewise, the committee was purposely not composed to conduct business case analyses related to the report recommendations. However, such analyses of selected recommendations are worthy of future consideration by the Air Force because they would provide insight on whether implementation of the selected recommendations would result in overall long-term reductions in sustainment costs.

ment (Recommendations 3-2, 4-1, and 4-6); (3) modest progress by the Air Force on governance (Recommendations 2-6, 4-2, 4-4, 4-7, 6-6, 6-7, 6-8, and 6-9); (4) lack of a single senior Air Force commander in charge of the entire sustainment enterprise (Recommendation 2-4); (5) metrics that do not adequately measure the key sustainment parameters (Recommendations 2-1, 2-5, 2-6, and 4-9); (6) a spare parts chain that encumbers effective and efficient maintenance production at all levels (Recommendation 4-3); and (7) ineffective processes to develop and transition technology for sustainment (Recommendations 5-1, 6-2, 6-3, and 6-4). The recommendations provided in this report have the potential to (1) focus management and leadership attention with commensurate authority; (2) relieve the stressors of organizational conflicts; and (3) improve efficiencies across the Air Force sustainment enterprise. Chapters 2 through 6 of this report present 45 related findings and 27 recommendations. Below are the key recommendations from Chapters 2 through 6.

KEY RECOMMENDATIONS

The Impact of Regulations, Policies, and Strategies on Sustainment

The lack of clearly defined sustainment goals affects the entire Air Force. Importantly, however, the Air Force can quickly develop a solution to this problem and then refine it over the longer term. The Air Force challenges in the sustainment process begin in the Air Force organizations—i.e., Assistant Secretary of the Air Force (Acquisition) (SAF/AQ), Assistant Secretary of the Air Force (Installations, Environment and Logistics) (SAF/IE), and Air Staff Headquarters (AF/A4/7)—that are responsible for the clarity of policy and process to the subordinate organizations. These headquarters offices should set the tone for Air Force sustainment. In the absence of well-founded policy and instructions, field-level commanders and directors take individual action to sustain their fleet; however, they deserve clear guidance and should be held accountable for execution.

Recommendation 2-1. The Air Force should establish sustainment goals that are specific and can be understood by all acquisition, contracting, engineering, and sustainment professionals. The Air Force should then track these goals and hold key individuals accountable for achieving them.

Recommendation 2-2. The Air Force should conduct a detailed holistic review of all appropriate sustainment policies and directives and build a complementary suite of processes and actions. With regard to the Title 10 mandates, the Air Force should take near- and long-term strategic actions to ensure maximum compliance. The Air Force should make long-term acquisition decisions and should aggressively pursue opportunities for Centers of Industrial and Technical Excellence.[8,9]

Recommendation 2-3. The Air Force should select and deploy either a single (preferred) or standard (by type of aircraft) collaborative engineering methodology to determine the processes, procedures, tasks, and frequency of maintenance actions.

Recommendation 2-4. The Air Force should consider formally designating a senior commander, such as the Air Force Materiel Command (AFMC) Commander, as the commander of the entire sustainment process, from concept phase through system retirement, with the responsibility to advise the SAF/IE, SAF/AQ, and AF/A4/7 on policy and then train, organize, equip, plan, and execute the Air Force's Integrated Life-Cycle Management (ILCM) processes.

Recommendation 2-5. The Air Force should develop and implement weapon system-level metrics that set aircraft availability levels and the cost of providing that availability, as well as identify who is responsible for attaining both. Furthermore, these measures should be at a level that reflects sustained imple-

---

[8]Quoting, in part, 10 USC § 2466 entitled "Limitations on the Performance of Depot-level Maintenance of Materiel": "(a) Percentage Limitation.— Not more than 50 percent of the funds made available in a fiscal year to a military department or a Defense Agency for depot-level maintenance and repair workload may be used to contract for the performance by non-Federal Government personnel of such workload for the military department or the Defense Agency. Any such funds that are not used for such a contract shall be used for the performance of depot-level maintenance and repair workload by employees of the Department of Defense." For additional information, see http://www.law.cornell.edu/uscode/uscode10/usc_sec_10_00002466----000-.html. Accessed July 8, 2011.

[9]Quoting, in part, 10 USC § 2474 entitled "Centers of Industrial and Technical Excellence: Designation: Public-Private Partnerships" and enacted November 18, 1997, by Public Law 105-58: "(1) The Secretary concerned, or the Secretary of Defense in the case of a Defense Agency, shall designate each depot-level activity of the military departments and the Defense Agencies (other than facilities approved for closure or major realignment under the Defense Base Closure and Realignment Act of 1990 (part A of title XXIX of Public Law 101-510; 10 U.S.C. 2687 note)) as a Center of Industrial and Technical Excellence in the recognized core competencies of the designee. (2) The Secretary of Defense shall establish a policy to encourage the Secretary of each military department and the head of each Defense Agency to reengineer industrial processes and adopt best-business practices at their Centers of Industrial and Technical Excellence in connection with their core competency requirements, so as to serve as recognized leaders in their core competencies throughout the Department of Defense and in the national technology and industrial base (as defined in section 2500(1) of this title [10 USCS § 2500(1)])." For additional information, see http://www.law.cornell.edu/uscode/html/uscode10/usc_sec_10_00002474----000-.html. Accessed July 19, 2011.

mentation of process improvement initiatives such as cost-reduction incentives and not just increasing sustainment costs driven by aircraft aging.

Recommendation 2-6. The Air Force should develop and direct a sustainment execution model for weapon systems and major weapon systems modifications that balances the needs of the individual weapon system with the performance goals and cost constraints of the overall sustainment enterprise.

Assessment of Current Sustainment Investments, Infrastructure, and Processes

As noted above, the Air Force currently spends billions of dollars per year to sustain legacy weapon systems. It is likely that the amount of money needed to sustain legacy weapon systems will increase over time as weapon systems become more complex. At the same time, current projections indicate that the Air Force budget will decrease over the remaining life of the Future Years Defense Program. These fiscal realities will lead to difficult choices for the Air Force in terms of sustaining the legacy fleet, recapitalizing the fleet, and modernizing the fleet. Although the latter two efforts are not addressed in this report, the Air Force sustainment enterprise does not, and will not, operate in a vacuum. The budget process must recognize the need for some up-front investments in the sustainment enterprise to increase efficiencies and effectiveness. Consequently, the sustainment of existing weapon systems will demand careful attention and frequent readiness, availability, and cost trade-offs.

Recommendation 3-1. The Air Force should continue funding depot plant and capital equipment and, at the same time, be guided by focused analyses to ensure that constrained funding is provided to the most critical sustainment needs to avoid future support impacts and to meet Title 10 considerations.

Assessment of Air Force Air Logistics Centers

Evaluation of the ALCs considers funds allocation, workforce, skill sets, and organizational structure. The flow of funds for depot maintenance and the Air Force's flying hour programs has been adequate; however, this might change in the near future. Even though the Air Force considers the ALCs to be the single most important and fundamental key to sustainment, their organizational structure is not adequately resourced, and the executive leadership does not have full command and control of the ALC enterprise. To a severe extent, the supply chain's ability to provide spare parts to the maintenance organizations is ineffective and disrupts depot maintenance and parts repair. There are mismatches between direct support to production activities and the growth of requirements at the production level. Importantly, a modern enterprise resource planning tool, although promised as "forthcoming," is not available, despite being desperately needed. Finally, and disappointingly, despite repeated requests for key measures of effectiveness and efficiency, the only well-defined metric supplied was for aircraft availability, and many more organizations than just the ALCs impact aircraft availability. The ALCs make the current situation work, but the full spectrum of resources needed to make an effective and efficient organization is not available.

Recommendation 4-1. The Air Force should establish streamlined command lines of accountability and authority to give the ALC commanders clear execution authorities to direct process improvements on assigned programs, maintenance activities, and supply support.

Recommendation 4-2. The Air Force should follow in a more timely manner the statutes that require the depot maintenance industrial workforce to be managed according to workload. The Air Force should also ensure that supporting organizations are staffed to support the industrial workloads and that flexible work rules are established to permit more workforce versatility.

Recommendation 4-4. The Air Force should continue its eLog 21 approach to sustainment improvement and should aggressively continue to pursue incremental fielding of the Expeditionary Combat Support System as an enterprise resource planning solution. Strong advocacy for this program should reside with the sustainment commander.[10]

Recommendation 4-5. The Air Force should focus the same, or arguably more, attention and investment as that given to equipment in the actual weapon system on the tools used for software maintenance. Maintaining currency between test laboratories and actual weapon systems is fundamental for dealing with timing, details of hardware interface behavior, and concurrency.

Recommendation 4-7. The Air Force should review its corporate oversight, management, and support of embedded systems software development and sustainment with foci on (1) greater risk identification and mitigation and (2) enabling Air Force corporate, as opposed to command-specific, decision making. In addition, given Air Force dependence on software to achieve mis-

[10]Based on detailed information provided to the committee by the Air Force, the committee concluded that there is a critical need for an Enterprise Resource Planning (ERP) system to help address Air Force sustainment issues. The Expeditionary Combat Support System is being developed with ERP capabilities and will be implemented to meet the Air Force's need. The committee did not attempt to evaluate the process that selected the specific ERP system or the contractor(s) involved in the development and implementation.

sion capabilities, the Air Force should strongly consider additional education on software sustainment for senior leaders.

Recommendation 4-9. The Air Force should develop key metrics for sustainment that flow to ALC commanders and that highlight the success or shortcomings of ALC activities, drive appropriate behavior for the workforce, and allow Air Force leadership to assess the health of the enterprise and the adequacy of resourcing for the sustainment process regardless of organizational affiliation.

Technology Development and Insertion for Sustainment

The Air Force's continued reliance on aging aircraft, such as the B-52, C-130H, A-10, F-16, and C-5B, which will exceed their originally designed life spans, will place an emphasis on the increasingly important role of new technologies related to materials, inspection systems, and vehicle health monitoring.[11] The Air Force Research Laboratory (AFRL), under the AFMC, is responsible for developing new technologies for Air Force weapon systems. AFRL can point to a long history of attention to sustainment and to many successful transitions of technology to industry and to the ALCs that increased aircraft availability and/or reduced maintenance costs. In recent years, support for sustainment-focused technology has waned because of increased attention to other technical priorities and opportunities, too frequent changes in strategy and process, and reductions in funding for sustainment technology, especially that for the transition process. There is capability within AFRL development programs to affect sustainment costs on existing weapon systems, and justification for increased AFRL investments and foci on sustainment.

At the time this report was being drafted, the Air Force began to implement high-level processes for technology development and insertion. Specifically, it was encouraging to learn that (1) sustainment is clearly identified in the recently released Air Force science and technology plan; (2) a new program element has been established for this area; and (3) the AFMC Commander has ownership of technology development and transition. In addition, as the Air Force moves toward the ILCM concept that is now beginning to appear in high-level plans and visions, it will be necessary to broaden the common understanding of sustainment technologies and to adequately support their inclusion, development, and transition into new weapon systems.

[11]Steven H. Walker, Deputy Assistant Secretary of the Air Force for Science, Technology, and Engineering. "Co-Sponsor Discussion of Study Background and Task." Presentation to the committee, October 20, 2010.

Recommendation 5-1. The Air Force should develop a "technology for sustainment" plan that identifies processes, technical agendas, workforce needs, and required funding resources. Such a plan should be imbedded within the overall ILCM strategy that is being developed by the Air Force.

Incorporating Sustainability into Future Designs

The Air Force and industry have unique capabilities for developing and maturing technology, as discussed in Chapters 5 and 6. When specifically focused, these capabilities can be applied to the incorporation of sustainment features into future designs to complement the traditional focus on system performance. The successful application of lessons learned from field experience is exemplified in some recent designs, and there are evolving human factors techniques that provide tools for integrating new maintenance functions and personnel capabilities. The experience gained from the deployment of more recent weapon systems containing special emphasis on low observable characteristics and significantly more use of software also provides a wealth of data for incorporation of sustainment capabilities into future designs. Discussions with contractors revealed that (1) in one case, much of the activity that resulted in the incorporation of sustainment features into the systems was initiated by the contractor and (2) in the other case, detailed sustainment requirements came from another Service. This again emphasizes the importance of having strong sustainment involvement in all phases of Air Force procurements, beginning with the concept and design phases. Incorporation of sustainment into future designs requires a strong partnership between government and industry.

Sustainment professionals have not been fully involved in the Defense Acquisition Management System (DAMS), from the Materiel Development Decision (MDD) to Production and Deployment (P&D), of recent weapon system programs. As a result, sustainment needs, planning, and costs have not been fully captured. A partnership between all sustainment stakeholders should be established in the initial requirements statements, should continue with the original equipment manufacturer at the earliest possible opportunity, and should drive future sustainment planning and technology insertion. It is important to emphasize the opportunities for government/contractor shared knowledge and alignment with Air Force enterprise programs in this partnership.

Recommendation 6-1. The Air Force should involve its sustainment professionals throughout the DAMS, from MDD to P&D, in all future weapon system development. Funding for sustainment planning and support should be given the same visibility as that for the development of capabilities and performance.

Recommendation 6-3. The Air Force should establish an institutionalized process for collecting and consistently incorporating desirable design features or applying lessons learned from legacy programs into the requirements for new systems or systems being modified in the support phase of the life cycle and into the internal procedures involved in sustaining these systems.

Recommendation 6-4. The Air Force should place more emphasis and implement additional training for acquisition professionals on the need for, the how to, and the pricing of proper data rights and domain knowledge related to the weapon system.

Recommendation 6-5. The Air Force should commit to establishing processes and resources that support consideration of a blended organic–contractor partnership early in the program life cycle and throughout the deployment and support phases.[12]

Recommendation 6-9. The Air Force should consider incorporating commercial-like engineering models and data collection and analysis techniques into the appropriate future platforms and contractually require that these efforts be compatible with Air Force data systems.

SUSTAINMENT, RECAPITALIZATION, AND MODERNIZATION: CAN THE AIR FORCE AFFORD TO DO EVERYTHING?

The short answer to this question is no. The high costs associated with Air Force weapon system sustainment will continue to directly impact the procurement of replacement and new systems unless significant numbers of legacy aircraft are retired.[13,14] To date, the Air Force sustainment enterprise has been largely successful in meeting the requirements of a 20-year-long, high-operational-tempo period. Going forward, however, high-level Air Force management, hopefully informed by

[12]A blended organic–contractor partnership, as specified in the contract award for a weapon system, assigns specific responsibilities for sustainment of a weapon system to both the industry contractor and the organic ALC.

[13]Steven Walker, Deputy Assistant Secretary of the Air Force for Science, Technology, and Engineering. 2011. "2012 Budget Request for DoD Science and Technology." Testimony to the House Armed Services Committee, March 1. Available at http://armedservices.house.gov/index.cfm/hearings?ContentRecord_id=b68a8831-3c7f-4ca2-a901-046b50eaa90d&Statement_id=7ea6c379-4499-4e88-b064-37e7b4d3f2d2&ContentType_id=14f995b9-dfa5-407a-9d35-56cc7152a7ed&Group_id=13e47ffa-0753-47a7-ad5e-1ba7592015c9&MonthDisplay=3&YearDisplay=201. Accessed April 22, 2011.

[14]Erin Conaton, Under Secretary of the United States Air Force. 2011. "The Future of the Air Force." Speech at the Center for Strategic and International Studies, Military Strategy Forum, March 31. Available at http://csis.org/event/future-air-force. Accessed April 24, 2011.

the recommendations in this report, will be called upon to effect organizational and operational improvements to make the Air Force sustainment enterprise more efficient. It cannot be known to what extent this report has been affected by not having insight into internal Air Force deliberations on the Defense Efficiencies Initiative. Even so, the findings and recommendations in this report stand on their own and reflect the synthesis of a voluminous amount of information.

Finally, What if the Air Force were to fully implement the recommendations in this report? *Many of the recommendations will focus management and leadership attention with commensurate authority, relieve the stress of organizational conflicts, and improve efficiencies across the board and can be done without investment other than that of personnel time.* In addition, the actions recommended throughout the report are expected to achieve long-term cost reductions that can help to produce a strong and affordable Air Force sustainment enterprise. Some recommendations, such as those for dramatically improving the supply of spare parts and implementing an enterprise resource planning system, will require up-front investments but can result in future cost savings. Other recommendations will require up-front expenditures—for example, funding for the most pressing sustainment needs of depot plant and capital equipment—to avoid future support impacts. In summary, this report presents recommendations aimed at improving the effectiveness and efficiency of the Air Force weapon system sustainment enterprise and at driving cost reductions over the long term.

1

Introduction

OVERVIEW

The ability of the United States Air Force (USAF) to keep its aircraft operating at an acceptable operational tempo, in wartime in and peacetime, has been vital to the Air Force since its inception. This is a much larger issue for the Air Force today, having effectively been at war for 20 years, with its legacy aircraft becoming increasingly more expensive to operate and maintain and with military budgets certain to further decrease. The enormously complex Air Force weapon system sustainment enterprise is currently constrained on many sides by laws, policies, regulations and procedures, relationships, and organizational issues emanating from Congress, the Department of Defense (DoD), and the Air Force itself. The difficulty of functioning in the midst of this complexity is compounded as the operational demands of weapon system sustainment and its growing cost collide with the realities of shrinking budgets in the years ahead. The Under Secretary of Defense for Acquisition, Technology and Logistics recently stated, "[M]ost of the money in the budget is spent on sustaining weapon systems that were procured in the past rather than on acquisition programs per se."[1] In May 2010, the DoD introduced the Defense Efficiencies Initiative, which seeks to "increase efficiencies,

[1]Ashton Carter, Under Secretary of Defense for Acquisition, Technology and Logistics. "Pentagon Efficiency Initiatives." Remarks given at the Heritage Foundation, Washington, DC, April 20, 2011. Available at http://www.heritage.org/Events/2011/04/Pentagon-Efficiency?query=Pentagon+Efficiency+Initiatives:+Are+They+Enough+to+Stave+Off+More+Defense+Cuts?. Accessed May 2, 2011.

reduce overhead costs, and eliminate redundant functions in order to improve the effectiveness of the DoD enterprise. This effort is focused on reprioritizing how DoD can use resources to more effectively support and *sustain* the force [emphasis added]."[2] As the overall Air Force budget decreases, funding for weapon system sustainment competes with funding for modernizing the Air Force. Against the back-drop of these stark realities, the Air Force requested the National Research Council (NRC) of the National Academies, under the auspices of the Air Force Studies Board, to conduct an in-depth assessment of current and future Air Force weapon system sustainment initiatives and recommend future courses of action for consideration by the Air Force.

COMMITTEE FORMATION AND TERMS OF REFERENCE

The National Academies approved the terms of reference (TOR) for the study in September 2009 (see Box 1-1), and this 15-month study was funded by the Air Force in July 2010. Committee members were selected for their backgrounds in academia, industry, and government. Two additional committee members with background and experience in weapon system software and Air Force financial management were added to the committee after the requirements and breadth of tasks were better understood.

STUDY APPROACH

Six full committee meetings of 2 to 3 days each were held approximately every month starting in October 2010. These full committee meetings were held at facilities operated by the National Academies and at Wright-Patterson Air Force Base. In addition, subgroups of the full committee visited the three Air Force Air Logistics Centers (ALCs), the Navy Fleet Readiness Center Southwest, and the Air Force Research Laboratory (AFRL).

The committee extensively relied on briefings by Air Force management teams in the acquisition and logistics functional disciplines on aircraft sustainment from the conceptual exploration of technology, to weapon system initial concepts, to weapon system fielding, to the sustainment phase, and eventually to system retirement. The committee conducted interviews and discussions with senior leaders, technical specialists, and managers, including those focused on (1) development of new capabilities; (2) acquisition activities for fielding new capabilities and systems; and (3) support of fielded systems including supply chain management, engineering, technical management, and overhaul and repair of current fleets. To assist its

[2]Department of Defense (DoD). 2010. Defense Efficiencies Initiative. Available at http://www.defense.gov/home/features/2010/0810_effinit/. Accessed April 21, 2011.

BOX 1-1
Terms of Reference

The NRC will

1. Assess current sustainment investments, infrastructure and processes for adequacy in sustaining aging legacy systems and their support equipment.
2. Determine if any modifications in policy are required, and, if so, identify them and make recommendations for changes in Air Force regulations, policies, and strategies to accomplish the sustainment goals of the Air Force.
3. Determine if any modifications in technology efforts are required, and, if so, identify them and make recommendations regarding the technology efforts that should be pursued, because they could make positive impacts on the sustainment of the current and future systems and equipment of the Air Force.
4. Determine if the Air Logistics Centers have the necessary resources (funding, manpower, skill sets, and technologies) and are equipped and organized to sustain legacy systems and equipment and the Air Force of tomorrow.
5. Identify and make recommendations regarding incorporating sustainability into future aircraft designs.[1]

[1]Many of the presenters to the committee agreed that a budget "train wreck" with respect to sustainment costs is looming. The TOR did not require the committee to undertake business case analyses related to the report recommendations. However, such analyses are worthy of future consideration by the Air Force because they would provide insight on whether implementation of the recommendations would result in overall long-term reductions in sustainment costs.

evaluation of Air Force sustainment activities, the committee also met with past and present naval aircraft support personnel, the Defense Logistics Agency, and industry experts associated with both military support and commercial aviation fleet management.

From the beginning of the study, the committee sought to understand what was meant in the terms of reference by the term "sustainment" and the phrase "sustainment goals of the Air Force." The committee also recognized the need to understand the "as is" conditions of sustainment support; comprehend the environments that the Air Force sustainment enterprise has faced in the past, faces now, and is likely to face in the future; and determine the Air Force's planning for future sustainment activities. In the DoD or the Joint Staff, sustainment with respect to weapon systems is not precisely defined. However, joint doctrine refers to "sustainment" planning for operations.

DEFINING SUSTAINMENT

Weapon System Sustainment in the Context of the Military Mission

The Joint Doctrine, Joint Force Employment, Planning for Joint Operations, J-7 Operational Plans and Interoperability Directorate defines "sustainment," in terms of "sustainment planning," as follows:[3]

> Sustainment planning is directed toward providing and maintaining levels of personnel, materiel, and consumables required to sustain the planned levels of combat activity for the estimated duration and at the desired level of intensity. It is the responsibility of the combatant commanders in close coordination with the Services and defense agencies. (p. 5)

The Defense Acquisition University (DAU) significantly expands the definition of sustainment as follows:

> the supportability of fielded systems and their subsequent life cycle product support—from initial procurement to supply chain management (including maintenance) to reutilization and disposal. It includes sustainment functions such as initial provisioning, cataloging, inventory management and warehousing, and depot and field level maintenance. Sustainment begins when any portion of the production quantity has been fielded for operational use. Sustainment includes assessment, execution and oversight of performance based logistics initiatives, including management of performance agreements with force and support providers; oversight of implementation of support systems integration strategies; application of diagnostics, prognostics, and other condition based maintenance techniques; coordination of logistics information technology and other enterprise integration efforts; implementation of logistics footprint reduction strategies; coordination of mission area integration; identification of technology insertion opportunities; identification of operations and support cost reduction opportunities and monitoring of key support metrics. [4]

DAU's definition of sustainment is broad in scope and nearly all encompassing. To best align the scope of the study, the committee examined other definitions that focus solely on the weapon system—for example, the Weapon System Acquisition Reform Product Support Analysis (2009), where the preferred term that most closely aligns with weapon system sustainment in the context of this study is "product support."

---

[3] Available at http://www.dtic.mil/doctrine/jrm/plans.pdf. Accessed April 22, 2011.

[4] Available at https://acc.dau.mil/CommunityBrowser.aspx?id=18073. Accessed May 1, 2011.

Weapon System Sustainment Defined for This Study

For this report, the committee adopts the following definition of weapon system or aircraft sustainment from the 2009 DoD report:[5]

> System sustainment is the package of support functions required to maintain the readiness and operational capability of weapon systems, subsystems, software, and support systems. It encompasses materiel management, distribution, technical data management, maintenance, training, cataloging, configuration management, engineering support, repair parts management, failure reporting and analysis, and reliability growth.

Past, Present, and Future Environments for Weapon System Sustainment

The committee hopes that this report is timely for the Air Force in light of the current environment of an uncertain world, more than 20 years of high-tempo operations, expanding global demands on the Air Force, and a high demand for continuous surveillance over current theaters of operations. Additionally, the Air Force as well as the entire DoD is under intense budgetary pressures, with rapidly escalating costs associated with weapon system sustainment due, in part, to significantly aging fleets and smaller numbers of newer fleets with features and capabilities that increase support costs. Simultaneously, there is a constant need to recapitalize the Air Force's aged fleet and to introduce new technology to provide the required level of deterrence and warfighting capability.

Military Operations

Since its formation 63 years ago, the Air Force has experienced variations in its aircraft readiness. Nevertheless, the Air Force has always made it a priority to keep its aircraft operating at acceptable rates of mission accomplishment and to be ready for any mission the nation's leaders direct. The Air Force has been on a wartime-like footing for the past 20 years, and its aircraft systems are aging and becoming increasingly more expensive to operate and maintain. The committee received no evidence to indicate that the demand for Air Force resources and the associated operational tempo will diminish in the near term. Additionally, the fleet mix has changed over the past 10 to 15 years from fleets that were typically "hardware oriented" with limited amounts of software to platforms entering the inventory

---

[5]DoD. 2009. DoD Weapon System Acquisition Reform Product Support Assessment. November. Washington, DC: Office of the Under Secretary of Defense for Acquisition, Technology and Logistics. Available at https://dap.dau.mil/career/log/blogs/archive/2010/01/28/implementation-of-dod-weapon-system-acquisition-reform-product-support-assessment-psa-recommendations.aspx. Accessed November 22, 2010.

today that are dependent on software for up to 80 percent of their functionality, with attendant upkeep costs.[6]

As many aircraft in the current fleet are extended beyond their planned lives, they are modified to extend their lives, resolve existing capability shortfalls, or improve hardware and software support situations. Yet, a newly modified platform, although achieving new levels of performance to meet military demands, is often still an aged platform. The committee saw evidence of updated aircraft systems experiencing aging issues in numerous presentations and at visits to the ALCs. For example, a newly modified C-5M aircraft needed depot maintenance actions to repair aircraft subsystem structure cracking that led to fuel leaks only months after the modifications were completed. Although modified aircraft are needed and are performing superbly, the basic airframe will require careful attention for the remainder of its useful life.

WEAPON SYSTEM SUSTAINMENT GOALS OF THE AIR FORCE

Even after much research and discussion, the committee was unable to identify the officially sanctioned sustainment goals for the Air Force. It was often stated during the course of the study that aircraft availability is the measure of merit. However, widely varying oral definitions for aircraft availability were provided. The committee closely examined Air Force Instruction 21-101 and Technical Order 00-20-2 and observed charts that show aircraft availability for various weapon systems.[7] The Air Force's sustainment goals are discussed in detail in Chapter 2 of this report.

Governance: Laws, Policies, Strategies, and Regulations

The enormously complex Air Force sustainment enterprise is currently constrained on many sides by laws, policies, regulations and procedures, and relationships emanating from the Congress, DoD, regulatory agencies such as the Occupational Safety and Health Administration and the Environmental Protection Agency, labor agreements, and the Air Force itself. In fact, during the course of the study, seldom was there a discussion that did not raise governance issues. These governances have been instituted to promote various standardized practices, facilitate the development of sustainment practices, and accommodate special interest

[6]Jack Ferguson. Crouching dragon, hidden software: Software in DOD weapon systems. 2001 IEEE Software 18(4):105-107.

[7]Lt Col Jeff Meserve, Chief, Congressionals, Studies and Analysis Branch, Directorate of Maintenance, DCS/Logistics, Installations and Mission Support. USAF Maintenance Metrics: Looking Forward with Aircraft Availability. Available at http://www.sae.org/events/dod/presentations/2007LtColJeffMeserve.pdf. Accessed April 29, 2011.

needs. Collectively they create a labyrinth of issues that detract from managing sustainment as a balanced enterprise with strong emphases on effectiveness for the warfighter and high efficiency for the taxpayer. In fact, a report commissioned by the National Defense Authorization Act of 2009 frequently points out that the governances are many, application is often not adhered to, and the results make sustainment measurements difficult.[8] There appears to be broad support among the Air Force acquisition and sustainment communities for a comprehensive and extensive review of these various governances. AFI 63-101[9] is an example of an excellent directive after which such a review could be fashioned. Also, as noted by the sustainment community, the Air Force has not delegated to a single office or command the authority to integrate both early acquisition direction on system sustainment practices as well as to control sustainment in the years of execution.

Relationships

Sustainment activities require significant coordination and communication across a myriad of functions and organizations. Sustainment is currently largely facilitated by interpersonal relationships rather than clear lines of authority. Although many sustainment activities and processes produce desired operational outcomes, many issues require great effort to just "make it happen."[10] A classic example is the supply chain fragmentation that occurred when the base realignment and closing (BRAC) actions of the 1990s, and particularly 2005, moved the procurement of components and parts to the Defense Logistics Agency. A significant portion of this review addresses the effects that past early-system configuration and programmatic decisions have had on operational sustainability policies and looks into process directives that would be helpful in facilitating a more effective means of putting process controls into practice.

Budget

The difficulty of managing and functioning in the midst of this complexity is compounded as the operational demands of sustainment and its growing cost collide with the realities of shrinking budgets. As the overall Air Force budget gets

---

[8]Logistics Management Institute (LMI). 2009. Future Capability of DoD Maintenance Depots: Interim Report. LG901M1. December. Mclean, Virginia: LMI. Available at http://armedservices.house.gov/index.cfm/files/serve?File_id=be97f304-3d15-4e96-bc24-689f8cb6c633. Accessed February 20, 2011.

[9]USAF. 2009. Acquisition and Sustainment Life Cycle Management. AFI 63-101. April 17. Available at http://www.e-publishing.af.mil/shared/media/epubs/AFI 63-101.pdf. Accessed December 14, 2010.

[10]A discussion of sustainment organizational authority and an example of the coordination efforts required is found in Chapter 4 and depicted in Figure 4-1.

smaller in real terms, funding for sustainment increasingly competes with funding to modernize the Air Force and maintain a strong and dedicated team. As discussed in Chapter 3, the real demands created by aging fleets, increasingly sophisticated systems with higher sustainment costs, and support concepts configured late in the lifecycle can drain the Air Force budget. There are many examples, such as the C-130 aircraft, an airlift workhorse, where the increased operational tempo has caused the scope of depot maintenance work to grow by 50 percent over the past several years.

REPORT ORGANIZATION

The complexity of the Air Force sustainment enterprise lends itself to a high degree of subject matter overlap between report chapters. This chapter provides a broad context in terms of historical factors related to sustainment and the important fact that a single agreed-upon definition of sustainment does not exist. Chapter 2 addresses element 2 of the TOR by analyzing statutes and DoD and Air Force policies and procedures that direct selected aspects of acquisition that influence sustainment, as well as governances that are directed at sustainment activities. The complex acquisition life cycle and actions affecting long-term sustainment activities are also discussed in Chapter 2. Chapter 3 covers element 1 of the TOR by exploring the broad aspects of the resources consumed by sustainment. There, historical Air Force budget and execution documents are examined and future trends extrapolated.[11]

Chapter 4 responds to element 4 of the TOR by analyzing the Air Logistics Centers in terms of resources, processes, and organization. Chapter 5 addresses element 3 of the TOR by examining the history of sustainment and the "art of the possible" in advancing technology into the current systems. High-payoff opportunities are currently rare because investment focused on technology insertion targeting sustainment is low and the "hurdle" rate for investment is extremely high. Chapter 6 deals with element 5 of the TOR by looking at opportunities and concepts that can be used to incorporate sustainability into future aircraft designs. There is some degree of overlap between Chapter 5 and Chapter 6 in terms of the application of

[11]The DoD budget for FY2011 was appropriated toward the end of the writing of this report, and the best available data were used for the analyses in the report. In addition, the President's FY2012 budget was released mid-way during the committee's deliberation and serves as a source document for analyzing future expected costs. Considerable time was devoted to trying to understand the description of the "efficiencies" outlined in the FY2010 and 2012 President's budgets and their actual net effect on sustainment. Importantly, the deliberations regarding Air Force sustainment efficiencies were service-sensitive, and the committee was unable to gain an understanding of related ramifications. The committee also had sessions with staff members of the House and Senate Armed Services Committee to gain their perspective on not only resource issues, but also policy considerations.

technology to sustainment of weapon systems; the principle difference is historical and current Air Force sustainment technology initiatives (Chapter 5) compared to how technology for sustainment may be considered in future weapon systems for operational capability and utility. Chapter 6 also offers a commercial model for aircraft engineering, maintenance, and sustainment for future consideration. Finally, the report is organized to provide the reader with a logical analysis of the issues from the macro aspect of governance to the details of sustaining new systems with technological innovation in the future. Findings and recommendations are embedded in the text of Chapters 2 through 6 after the supporting evidence.

2

Review of the Impact of Regulations, Policies, and Strategies on Sustainment

INTRODUCTION

This chapter addresses element 2 of the terms of reference (TOR), that is, "Determine if any modifications in policy are required, and, if so, identify them and make recommendations for changes in Air Force regulations, policies, and strategies to accomplish the sustainment goals of the Air Force." The committee extensively examined the regulations, policies, and strategies that affect Air Force weapon system sustainment. The sustainment process actually should begin on the day when a warfighter states, "We need a…product" and should continue until the product that evolves from that need is retired to a defense recycling center or, in the case of a weapon system, a reclamation center—often 50 years later. It is estimated that 65 percent or more of the life-cycle costs relate to supporting rather than acquiring the system. [1] Sustainment concepts should be considered alongside of required capabilities at the beginning of the acquisition cycle, but this has not always happened. Nevertheless, progress toward this goal is being made as evidenced by new congressional interest and legislation, updated Department of Defense (DoD) policies, instructions, and reports, and substantial steps by the Air Force to incorporate an enterprise perspective into the acquisition and sustainment

[1] Sue Lumpkins, Principal Deputy Assistant Secretary of the Air Force for Installations, Environment and Logistics. "Air Force Studies Board Sustainment Study." Presentation to the committee, October 20, 2010.

effort. Although significant work remains, there is recognition within the Air Force that sustainment is important, complex, and expensive.

AIR FORCE SUSTAINMENT GOALS

Element 2 of the TOR states, in part, "to accomplish the sustainment goals of the Air Force." The committee maintained a constant focus on the Air Force sustainment goals, not only for the purpose of addressing them in this report, but also because *goals drive organizations.* When a goal is generally understood, then there are more likely to be a common purpose and higher probability of success. In its numerous discussions with Air Force personnel, the committee posed the following questions:

- What are the sustainment goals of the Air Force?
- Who sets the goals?
- How are the goals established?
- Who knows the goals?
- How are the goals tracked and to whom are they reported?

The answers were interesting to say the least. In fact, the majority of respondents were unable to respond in detail. Many talked about aircraft availability (AA), but, as discussed later in this chapter and in Chapter 4, AA is a parameter that has many complex elements.[2] Overall, who is in charge of achieving AA is in question. In its search for key sustainment goals, the committee became aware of a measure for improvement over time during a briefing on Expeditionary Logistics for the 21st Century (eLog21), which is a transformational Air Force campaign to drive supply chain improvements. The essence of a goal is represented by the last two bullets in Figure 2-1.

Despite what many people appear to believe, the goals for eLog21 do not also serve as the goals for Air Force weapon system sustainment. AA is far more widely reported than cost parameters as the goal. Committee members engaged various government and industry officials in extensive discussions regarding methodologies for assessing and calculating AA as well as which officials are accountable for AA. Unfortunately, the calculation responsibilities were better defined than the target's ownership. In fact, after thoroughly investigating AA, the committee concluded that AA ownership is vague at best and that, because there are so many "cooks in the AA kitchen," no one can be held truly accountable. The report's discussion of

[2]See, for example, the "quad chart" from: Ashton Carter, Under Secretary of Defense for Acquisition Technology, and Logistics. "Strengthened Sustainment Governance for Acquisition Program Reviews." Memorandum for Secretaries of the Military Departments. April 5, 2010.

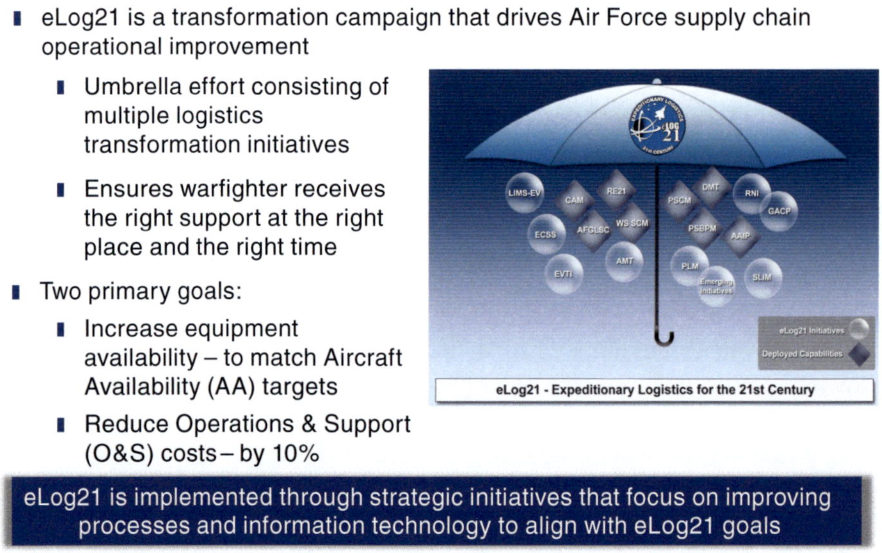

FIGURE 2-1
eLog21 defined. SOURCE: Grover L. Dunn, Director of Transformation, Deputy Chief of Staff for Logistics, Installations and Mission Support, Headquarters U.S. Air Force. "Expeditionary Logistics for the 21st Century (eLog21)." Presentation to the committee, January 17, 2011.

this issue is brief, because the committee found no well-established, documented, top-level sustainment goals with clear ownership for the Air Force.

Finding 2-1. The Air Force does not have (a) consistent, widely understood goal(s) for aircraft sustainment.

Recommendation 2-1. The Air Force should establish sustainment goals that are specific and can be understood by all acquisition, contracting, engineering, and sustainment professionals. The Air Force should then track these goals and hold key individuals accountable for achieving them.

STATUTES, REGULATIONS, AND POLICIES THAT IMPACT AIR FORCE SUSTAINMENT

Overview

The complexity of the Air Force sustainment enterprise is reflected in the myriad of statutes, regulations, and policies that the Air Force must consider when performing sustainment tasks. These include legislation that specifies the definition of depot maintenance (10 U.S.C. § 2460), types of work for the maintenance depots and core considerations (10 U.S.C. § 2464), amounts of work to be conducted in the depots, i.e., 50/50 accounting (10 U.S.C. § 2466), and funding levels to be spent on depot maintenance Capital Investment and Process Improvement, e.g., the 6 percent rule (10 U.S.C. § 2476); regulations derived from the mid-1980s' Acquisition Reform Act and the Weapon System Acquisition Reform Act, which require major acquisition programs to have a weapon system Product Support Manager (PSM) (10 U.S.C. § 805). Additionally, legislation that affects public-private partnerships (10 U.S.C. § 2474) allows for the creation of Centers of Industrial and Technical Excellence that directly affect depot maintenance activities. These centers are designated by the Service Secretaries and offer the industrial complex the opportunity to partner with commercial industry to gain technological investment and better support future technologies. Regulations of state and federal agencies, such as the Environmental Protection Agency, Occupational Safety and Health Administration, and the Federal Aviation Administration also affect all levels of the sustainment process, including military operations around the globe. These regulations are often augmented by DoD and Air Force instructions and policies designed to facilitate interdepartmental and intra-Air Force activities and imposed on every level of maintenance, supply, and transportation operations within the Air Force. Lastly, labor agreements impact many parts of the sustainment process.

Figure 2-2 highlights the complexity of these statutes, regulations, and policies in the context of the Air Force sustainment enterprise. The realities listed in Figure 2-2 are accepted at face value as policies to be updated over time; however, they address a myriad of influences and players—statutes, policies, programs, strategies, organizations, and more. For example, the 3-year-old Air Force Global Logistics Support Center is responsible for the full spectrum of the Air Force spare parts. (Policy issues remain in the supply arena, and Chapter 4 includes significant coverage of spare parts support.)

With Centralized Asset Management, operations and maintenance funds are consolidated during the execution year to enable the Air Force as a whole to sustain the forces. Core/50/50 refers to legislation that directly affects acquisition strategies and sustainment over the weapon system life cycle. The appointment of a PSM to major programs is dictated by Section 805 of the 2010 NDAA (Weapon System Reform Act above). In fact, every "bullet" in the figure has a profound implication for sustainment. The point of the chart extends beyond the fact that the policy must be updated. *The entire acquisition and sustainment environment is tightly woven and cannot be separated into individual elements without considering the effects on the whole.* Even more importantly, synchronization of all of these bullets—and many other "topical bullets" like these—is a herculean task. Without well-defined

- Airworthiness
- Centralized Asset Management
- Core/50/50
- Data Rights
- Depot Source of Repair
- Global Logistics Support Center
- Product Support Manager
- Public/Private Partnerships
- Performance-Based Life Cycle
- Program Support Reviews
- Rapid Response Process
- Reliability, Availability, and Maintainability
- System Integrity
- Sustaining Industrial Capabilities
- Sustainment Quad Chart
- Sustainment Readiness Levels
- Transition Support Plan

FIGURE 2-2
Sustainment policy updates to be addressed in Air Force policy. SOURCE: Blaise J. Durante, Deputy Assistant Secretary for Acquisition Integration, Office of the Assistant Secretary of the Air Force for Acquisition. "Budgeting Considerations Related to Sustainment." Presentation to the committee, October 21, 2010.

regulations and a key commander in charge, long-term positive effects are unlikely to be achieved.

Statutes and Regulations

As stated above, numerous statutes and regulations influence the sustainment process. Throughout the study, the committee reviewed these statutes and regulations and assessed their impacts. Although the statutes principally affect depot maintenance operations and processes, there is without question a number of regulations and policies, including policies from the 1980s' acquisition reform, that drive DoD and Air Force sustainment practices. In the acquisition area, the Service Acquisition Executive (SAE), Program Executive (PEO), and Single or System Program Manager (SPM) have primacy for decisions on a weapon system during the acquisition process. In today's operations, it is rare that a major Air Force weapon system program does not have some type of ongoing acquisition event. Consequently, the acquisition community has a constant role in the sustainment process. The 2010 Weapon System Acquisition Reform Act directs major programs to have a PSM. By enacting this legislation, Congress has emphasized the importance of thoroughly considering sustainment in major acquisition programs just as it emphasized through other legislation the importance of the nation's industrial depot maintenance capabilities.

Legislation affects depot maintenance operations to a large extent, beginning with a definition of depot maintenance by 10 U.S.C. § 2460. The statutory definition of depot maintenance is, at best, ambiguous and subject to interpretation.

In fact, although the definition seems to be widely accepted by most informed personnel, a review of the literature revealed that the Army, Navy, and Air Force take different interpretations of depot maintenance when conducting analyses and reporting. Different interpretations by different services directly impact the development of Core capability requirements and sustainment of workloads, as well as the military's 50/50 calculations.[3] Although the impact on the immediate sustainment process is minimal, it detracts from clear policy execution, which leads to debate on sustainment concepts for Air Force weapon systems. The general sense of those outside the Air Force is that the DoD as a whole must standardize what is and what is not depot maintenance and then report on activities accordingly.[4]

Because there is debate across the DoD about what is and what is not depot maintenance, it is worth asking the following questions: Does determining what is or is not depot maintenance really make a difference to the outcome of the sustainment process? Doesn't the work need to be done regardless of what it is called? The answer to the first question is another source of great debate. Without a firm and universally accepted definition of depot maintenance, program managers will make decisions that best optimize their own programs. The answer to the second question is "Yes, however..." Including or not including the various types of work, by definition, affects Core workload determinations, which, in turn, affects maintenance capability and 50/50 work allocations and results.[5] 10 U.S.C. § 2464, Core Logistics Capability establishes

> a requirement to maintain a core depot maintenance capability that is government-owned and government-operated (including government personnel) to ensure a ready and controlled source of technical competence and resources necessary to ensure effective and timely response to mobilizations, national defense contingency situations, and other emergency requirements.[6]

Core requirement determinations are exceedingly complex. The Air Force performs extensive analyses to determine the wartime taskings for its various aircraft,

---

[3]Logistics Management Institute (LMI). 2011. Future Capability of DoD Maintenance Depots. LG901M2. February. McLean, Virginia: LMI. Section 4, p. 4-2. Available at http://armedservices.house.gov/index.cfm/files/serve?File_id=394b31e6-4adc-47ca-a6f5-21547f0751fa. Accessed February 20, 2011.

[4]LMI. 2011. Future Capability of DoD Maintenance Depots. LG901M2. February. McLean, Virginia: LMI. Section 1, pp. 1-18. Available at http://armedservices.house.gov/index.cfm/files/serve?File_id=394b31e6-4adc-47ca-a6f5-21547f0751fa. Accessed February 20, 2011.

[5]LMI. 2011. Future Capability of DoD Maintenance Depots. LG901M2. February. McLean, Virginia: LMI. Section 4, pp. 1-19. Available at http://armedservices.house.gov/index.cfm/files/serve?File_id=394b31e6-4adc-47ca-a6f5-21547f0751fa. Accessed February 20, 2011.

[6]General Military Law, U.S. Code, Title 10, Section 2472 (February 10, 2010).

the wartime depot maintenance requirements generated by the wartime surge, and then the representative types and mix of work that the depots must supply to meet the wartime surge needs. Then, applying the 50/50 rule (10 U.S.C. § 2466, Distribution of Workload), the Air Force determines the portion of depot maintenance work that can be performed outside of the organic depots, that is, by contractors.

The summary effect of these two statutes was widely discussed by many of the speakers listed in Appendix B. From a process standpoint, many speakers pointed to the effort that goes into determining Core and 50/50 management and 50/50 reporting. There is, however, a much more insidious issue: the primacy of well-intentioned PEOs and SPMs who tend to make sustainment decisions that optimize individual platforms rather than take a more enterprise perspective. For years, this primacy drove decisions that went outside the Core workload process and produced adverse 50/50 spreads. In many cases, short-term economics also drove decisions: it costs program dollars to equip depots to maintain new systems, and such investments can be reduced or eliminated by deciding to use a variety of partnering or full contractor support. This situation was further exacerbated by a lack of checks and balances over decisions that might adversely affect the Air Force sustainment environment. The Air Force Materiel Command (AFMC) Commander acknowledged that he was responsible for sustainment during the year of execution,[7] but he did not identify who was responsible for enterprise-level decisions during the planning process. It appears that the inputs prior to the year of execution are not significant, but once in the year of execution, the AFMC Commander must determine how the pieces come together and how to finish the year in compliance with all of the statutes, policies, and directives.

These short-term or longer-term effects place the acquisition community and the sustainment community at great odds. Compliance with statutes conflicts with the desire to bring weapon system economics into balance and to simultaneously support the weapon system at the optimal level. The committee found that there has been no strong arbitrator for sustainment, policies on who can influence or reshape sustainment decisions have been unclear, and SPMs and PEOs have had decision-making authority. Unless an individual of great stature makes a compelling case to the SAE, the decision stands. Consequently, the Air Force has been compelled to closely analyze Core determinations to ensure that it has the capability and capacity to meet wartime needs, and today the Air Force is repeatedly on the verge of breaking the 50/50 rule. As pointed out by the LMI, the services react not to Core analysis and determination, but to the 50/50 re-

[7]General Donald J. Hoffman, Commander, Air Force Materiel Command, Wright-Patterson Air Force Base. Personal communication to the committee, December 9, 2010.

porting, which must be certified to the Office of the Secretary of Defense (OSD) and reported to the Congress:

> As currently constructed and used, core is viewed more as a reporting requirement than a management tool. More than one senior military official stated that considerable attention was paid to the requirements to comply with 10 U.S.C. §2466, the socalled 50/50 requirement, because the results were reported to Congress, and they anticipated adverse actions for failure to comply. In contrast, core requirements received no visibility above OSD.[8]

When a violation of the 50/50 rule occurs, the Secretary of Defense must grant a waiver on the basis of a national security determination. The committee was repeatedly told during its visits to Headquarters AFMC, Headquarters Aeronautical Systems Center, and the Air Logistics Centers (ALCs) of the extraordinary lengths the Air Force has taken to avoid violations of the 50/50 rule. The consequences of these actions on finances, readiness, and ALC work loading are discussed in Chapter 4.

The committee discussed at length all of the statutes that affect sustainment and particularly depot maintenance and determined that the Air Force at large has not systematically dealt with Core considerations, the 50/50 workload mix, and other statutory conditions. After meeting with three congressional staffers, the committee doubts that these statutes will change substantially.[9] By dealing with these statutes and leveraging U.S.C. 10 § 2474 (Centers of Industrial and Technical Excellence) to the extent possible; the Air Force can reduce some of the stressors on the 50/50 calculation. Additionally, instilling discipline into the overall sustainment determination process could help to eliminate the Core and 50/50 issues.

As dynamic as the legislation is on depot maintenance and acquisition efforts, the committee found little to no statutory language about managing materiel or the spare parts piece of the supply chain. However, the committee did find appropriate DoD instructions and policies on spare parts management and ample Air Force guidance and direction on material management. As Figure 2-3 indicates, the material management policies and instructions are extensive and considered satisfactory from the process and effect perspectives. However, the committee

---

[8]LMI. 2009. Future Capability of DoD Maintenance Depots: Interim Report. LG901M1. December. McLean, Virginia: LMI. Section 4, pp. 4-12. Available at http://armedservices.house.gov/index.cfm/files/serve?File_id=be97f304-3d15-4e96-bc24-689f8cb6c633. Accessed February 20, 2011.

[9]Peter Levine, Senate Armed Services Committee, General Counsel, Readiness & Management Support Subcommittee, Majority Lead; Lynn Williams, House Armed Services Committee, Readiness Subcommittee, Majority Lead; and Vickie Plunkett, House Armed Services Committee, Readiness Subcommittee, Minority Lead. Presentation to the committee, February 16, 2011.

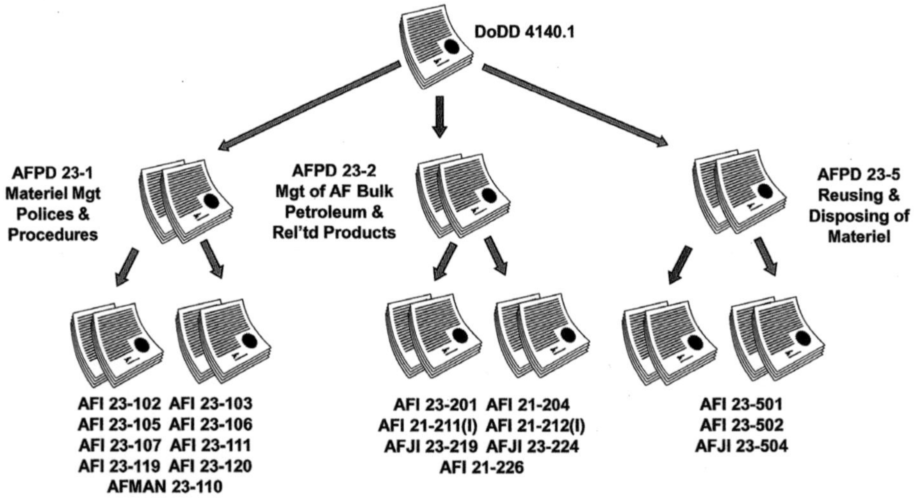

FIGURE 2-3
Material management. SOURCE: Major General Judith A. Fedder, Director of Logistics, Deputy Chief of Staff for Logistics, Installations and Mission Support, Headquarters United States Air Force. "AF Sustainment Study—Product Support, Material Management, Transportation." Presentation to the committee, February 14, 2011.

repeatedly observed material shortages to depot repair lines and to a lesser extent field activities and was informed by depot craftsmen of decisions made for some systems to switch from contractor support to organic support. Although the policies may be acceptable, there clearly are some execution issues with regard to material management. Chapter 4 discusses parts availability.

Finding 2-2. Over the years, Title 10 of the U.S. Code, DoD, and the Air Force have issued singular "directives" that have resulted in confusion, inefficiencies, and lack of accountability in the sustainment process. Title 10 is the "law of the land," and because it is unlikely that any "sea-state-change" will occur, the Air Force must come to grips with its provisions.

Recommendation 2-2. The Air Force should conduct a detailed holistic review of all appropriate sustainment policies and directives and build a complementary suite of processes and actions. With regard to the Title 10 mandates, the Air Force should take near- and long-term strategic actions to ensure maximum compliance. The Air Force should make long-term acquisition decisions and should aggressively pursue opportunities for Centers of Industrial and Technical Excellence.[10,11]

ESTABLISHING AIR FORCE SUSTAINMENT POLICIES

Sustainment must be thought of as a process and operated from a "systems" perspective. Air Force sustainment policies should establish a systems approach from concept to retirement and should be based on the following:

1. Engineering-based decisions with regard to the processes used to sustain an organization's operational systems;
2. A comprehensive approach to gathering and analyzing data;
3. A well-defined governance structure designed to ensure compliance with the directives prescribing the sustainment processes;
4. Organizational structures optimized and resourced to ensure proper leadership, training, force development, and execution of the sustainment processes;
5. A culture of collaboration whereby each of the many acquisition and sustainment functions, processes, and procedures is designed with an understanding of how individual actions affect the entire Air Force warfighting enterprise; and

---

[10]Quoting, in part, 10 USC § 2466 entitled "Limitations on the Performance of Depot-level Maintenance of Materiel": "(a) Percentage Limitation.— Not more than 50 percent of the funds made available in a fiscal year to a military department or a Defense Agency for depot-level maintenance and repair workload may be used to contract for the performance by non-Federal Government personnel of such workload for the military department or the Defense Agency. Any such funds that are not used for such a contract shall be used for the performance of depot-level maintenance and repair workload by employees of the Department of Defense." For additional information, see http://www.law.cornell.edu/uscode/uscode10/usc_sec_10_00002466----000-.html. Accessed July 8, 2011.

[11]Quoting, in part, 10 USC § 2474 entitled "Centers of Industrial and Technical Excellence: Designation: Public-Private Partnerships" and enacted November 18, 1997, by Public Law 105-58: "(1) The Secretary concerned, or the Secretary of Defense in the case of a Defense Agency, shall designate each depot-level activity of the military departments and the Defense Agencies (other than facilities approved for closure or major realignment under the Defense Base Closure and Realignment Act of 1990 (part A of title XXIX of Public Law 101-510; 10 U.S.C. 2687 note)) as a Center of Industrial and Technical Excellence in the recognized core competencies of the designee. (2) The Secretary of Defense shall establish a policy to encourage the Secretary of each military department and the head of each Defense Agency to reengineer industrial processes and adopt best-business practices at their Centers of Industrial and Technical Excellence in connection with their core competency requirements, so as to serve as recognized leaders in their core competencies throughout the Department of Defense and in the national technology and industrial base (as defined in section 2500(1) of this title [10 USCS § 2500(1)])." For additional information, see http://www.law.cornell.edu/uscode/html/uscode10/usc_sec_10_00002474----000-.html. Accessed July 19, 2011.

6. Outcome-based metrics designed to ensure appropriate availability of the operational systems to ensure mission success.

Engineering-based Decisions

As shown in Figure 2-4, the Air Force's senior sustainment leadership has a fundamental understanding of the challenges presented by the six principles. As a starting point, "Technical decisions drive everything that logisticians do."[12] Sustainment actions proceed on the basis of engineering determinations.

When dealing with complex systems with very distinct technology advances, the DoD must be very deliberate in defining the processes it uses to determine the failure modes likely to occur in systems that are unique, highly advanced, and based on highly complex integration of leading-edge technologies. Neither the DoD nor the Air Force has adopted a single, integrated, fact-based, analysis-derived process for determining the actions necessary to ensure the integrity of operational systems throughout their life cycles even though the Air Force indicates that it has employed the Reliability Centered Maintenance (RCM) and Failure Modes, Effects and Criticality Analysis (FMECA) processes since the late 1970s.

When the Air Force engaged in major outsourcing activities during the 1990s, it did not always stipulate a standardized maintenance management concept to be used by the prospective competitors. The contractors developed individual contract processes and procedures that may or may not have followed Air Force traditional practices. Likewise, the processes to determine the tasks, procedures, and frequencies of maintenance actions to operate and sustain systems vary not only between the ALCs, but also between similar systems within the same ALC. As a result, the Air Force uses several engineering-based processes to sustain its aircraft fleet. Some are based on RCM and FMECA, some are based on the processes developed by the original equipment manufacturer (OEM), some are based on an airline-developed process known as Maintenance Steering Group Three (MSG-3), and some are based on what are best described as hybrids.[13,14]

As Figure 2-4 outlines, there is good life-cycle management engineering and policy, but the implementation is lacking. The results, highlighted in Figure 2-4,

[12]Grover L. Dunn, Director of Transformation, Deputy Chief of Staff for Logistics, Installations and Mission Support, Headquarters U.S. Air Force. Personal communication with committee members, Pentagon, Arlington, Virginia, February 14, 2011.

[13]United States Air Force (USAF). 1980. Military Standard - Procedures for Performing a Failure Mode, Effects and Criticality Analysis MIL-STD-1629A. AMSC N3074. November 24. Available at http://www.everyspec.com/MIL-STD/MIL-STD+(1600+-+1699)/MIL_STD_1629A_1556/. Accessed May 2, 2011.

[14]Jim Gray, Headquarters Air Force Materiel Command/Engineering Directorate, Executive Officer. Personal communication to the committee, April 11, 2011.

- Technical decisions drive everything that logisticians do
  - Technical input drives planning and execution of supply and maintenance
- Life cycle system engineering and OSS&E policy are very good but sustainment implementation is not
  - Nonstandard, reactive, non-value-added processes andsystems
  - Missing/incomplete data to support downstream engineering
  - Disparate performance data andfew predictive "integrity" tools
  - Lack of "as fielded" configuration management discipline
- Significant disconnects between engineering/technical and planning/production processes and systems—e.g., BOMs
- The result is reduced availability and increased cost

Must Move from Reactive to Proactive Management

FIGURE 2-4
Product engineering: problem statement and gaps. SOURCE: Grover L. Dunn, Director of Transformation, Deputy Chief of Staff for Logistics, Installations and Mission Support, Headquarters U.S. Air Force. "Expeditionary Logistics for the 21st Century (eLog21)." Presentation to the committee, January 17, 2011.

are non-standard processes and ultimately "the results are reduced availability and increased cost." During the committee's tours of the ALCs and discussions with selected personnel at Wright-Patterson Air Force Base, the non-standard approaches were highlighted. Ultimately, the lack of a single, disciplined process by which the Air Force determines its maintenance actions puts its engineering efforts into a reactive posture, which is particularly troubling given the aging fleet the Air Force must operate now and into the future.[15]

> **Finding 2-3.** The Air Force has neither a single collaborative, engineering process nor a standard process for types of aircraft to determine the appropriate

[15] During the study, committee members with extensive experience in sustaining civil aviation systems cited the increased efficiency of operations and effectiveness of problem resolution as a direct result of having a single collaborative, engineering methodology. In addition, committee members from the manufacturing sector, with experience in both civil and military equipment, noted the effectiveness of the Federal Aviation Administration-enforced methodology in assuring timely fleet alerts and modifications, in contrast to a less collaborative approach currently used by the Air Force.

tasks, procedures, and frequency of maintenance actions to operate and sustain its operational systems.

Finding 2-4. The Air Force does not aggressively pursue standardized engineering programs that could improve the maintenance programs for the aircraft, accelerate process improvements, and increase aircraft availability across the service.

Recommendation 2-3. The Air Force should select and deploy either a single (preferred) or standard (by type of aircraft) collaborative engineering methodology to determine the processes, procedures, tasks, and frequency of maintenance actions.

Comprehensive Approach to Gathering and Analyzing Data

Over the past four decades, the DoD and Air Force focus has shifted with regard to the basic concept used to describe how the services will gather and use data to orchestrate the sustainment processes. Although the Air Force has varied in its engineering approach to sustainment, and more specifically in its maintenance, processes, and procedures, it has developed a well-designed series of data gathering and documentation endeavors to allow for a more predictive approach in some of its more critical subsystems. For example, the Aircraft Structural Integrity Program (ASIP), the PACER CENTURY effort on early F100-PW-100 engines, and the Joint Oil Analysis Program were all designed to provide current data on the condition of an aircraft's structural integrity, potential maintenance needs or failure of aircraft engines, or the potential failure of oil wetted systems in engines, respectively. However, there are shortcomings; for example, not all procured aircraft are equipped with a full ASIP capability. Data are captured on approximately one of six aircraft, and the condition of the remainder of the fleet is determined by extrapolations and like-mission scenarios.

Because the above endeavors showed some success and because better data automation and information management systems, such as the Core Automated Maintenance System (CAMS)/Reliability Maintenance Information System (REMIS), were implemented in the 1980s and early 1990s, the Air Force became better able to understand some aspects of a selected fleet's condition. Over time the Air Force developed new capabilities based on not only the theoretical aspects from tabletop RCM and FMECA studies but also actual data. These new capabilities were more scientifically predictive and were known as Condition-Based Maintenance (CBM).[16] The CBM concept, combined with better data efforts, enabled

[16] Chapter 5 contains additional discussion on CBM.

the Air Force to see its way to other new concepts such as Autonomic Logistics Information Systems. From automated and enhanced data collection, the sustainment community will realize a growth in opportunities to better understand and determine appropriate maintenance actions and consequently to greatly improve the effectiveness and efficiency of its enterprise. Unfortunately, these opportunities have not been capitalized on, and it will be some time before breakthrough efforts are fielded. Until then, the results will continue to be "best possible estimate engineering" from the best available data. Although the data have not improved substantially over the past few years, the DoD continues to espouse programs that demand even more data.

Currently, the DoD encourages the sustainment community to move toward a CBM+ concept that, similar to CBM, takes advantage of increasingly available technologies in the form of embedded sensors, mission scenarios, and external events, which can help to reduce maintenance actions to only those necessary to prevent a system, subsystem, or part failure. In theory, such a concept makes sense, but only when accompanied by rigorous, engineering-based analyses of the criticality of potential failures. The analysis must consider the consequences of predicting a failure and taking preventative maintenance actions versus performing maintenance actions after a failure occurs. Furthermore, CBM+ requires an investment to develop the specific technology for the specific platform and then install the capability across the fleet.

As outlined above, the processes used to determine appropriate maintenance and sustainment actions are engineering-based, have evolved over the past four decades, and are fully integrated in the commercial aviation world. However, the Air Force has not been able to apply the latest techniques to the legacy aircraft. In an environment of increasing sustainment costs, as discussed in Chapter 3, the Air Force's inability to advance to a data-based engineering solution for sustainment activities is of particular concern.

Current Governance Structure for Air Force Sustainment

The governance of the sustainment enterprise is complex and involves many organizations, all of which have vested interests in the processes, procedures, and outcomes. The centerpiece of the sustainment strategy is founded on the concept of an Integrated Life-Cycle Management (ILCM) enterprise. Laudably, the Air Force acquisition and sustainment communities are working together and embrace the concept shown in Figure 2-5.[17]

[17]Blaise J. Durante, Deputy Assistant Secretary for Acquisition Integration, Office of the Assistant Secretary of the Air Force for Acquisition. "Budgeting Considerations Relating to Sustainment." Presentation to the committee, October 20, 2010.

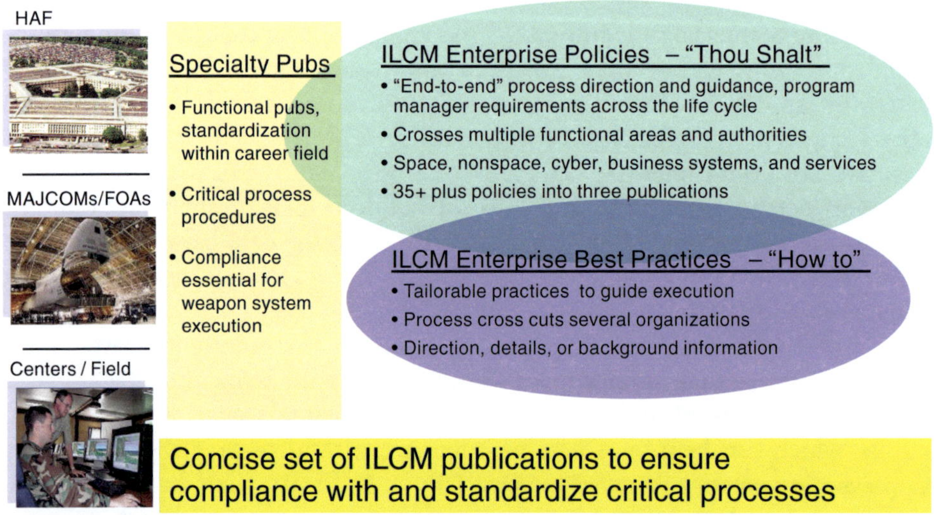

FIGURE 2-5
The concept behind the Integrated Life-Cycle Management (ILCM) enterprise. SOURCE: Blaise J. Durante, Deputy Assistant Secretary for Acquisition Integration, Office of the Assistant Secretary of the Air Force for Acquisition. "Budgeting Considerations Related to Sustainment." Presentation to the committee, October 21, 2010.

The integrated life-cycle concept is being codified by a single Air Force Instruction (AFI) 63-101. To properly define this effort, the Air Force has begun the process of combining into a single document the various functional directives previously designed to control the governance and policies of the various communities contributing to the sustainment processes. The introduction to this document states:

> This instruction must be used in conjunction with Air Force Instruction (AFI) 10-601, *Capabilities-Based Requirements Development*, AFI 99-103, *Capabilities Based Test and Evaluation*, AFI 63-1201, *Life Cycle Systems Engineering*, and AFI 20-101, *Logistics Strategic Planning Procedures*, to provide an integrated framework for the implementation of ILCM.[18]

Figure 2-6 shows the Air Force product support on ILCM. AFI 63-101 clearly states the need to integrate the efforts of the various functional elements that contribute to the ILCM enterprise. As the above quote shows, ILCM encompasses requirements, test, systems engineering, and sustainment, and, by its very framework, acquisition. The ILCM enterprise's primary mission is to provide seamless

[18]USAF. 2009. Acquisition and Sustainment Life Cycle Management. April 8. Available at http://www.e-publishing.af.mil/shared/media/epubs/AFI63-101.pdf. Accessed August 18, 2011.

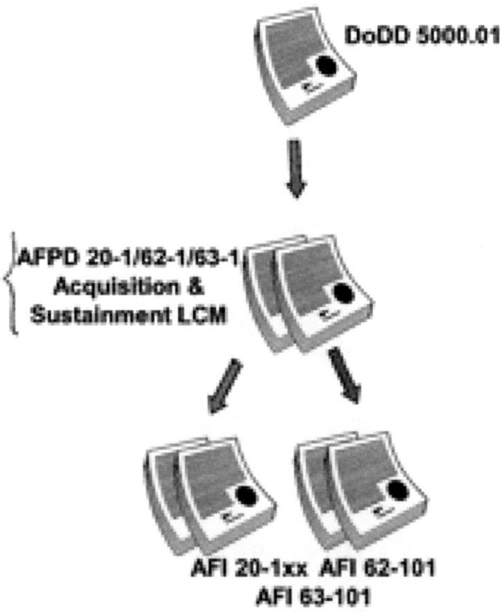

FIGURE 2-6
Product support. SOURCE: Major General Judith A. Fedder, Director of Logistics, Deputy Chief of Staff for Logistics, Installations and Mission Support, Headquarters United States Air Force. "AF Sustainment Study—Product Support, Material Management, Transportation." Presentation to the committee, February 14, 2011.

governance, policy, transparency, and integration of all aspects of weapon system acquisition and sustainment. Figure 2-7 demonstrates the effort to consolidate the various policies into one that is applicable to all communities. No longer will one community be able to assert that what is in another community's instructions does not apply to its respective effort. This is a very positive step; however, many remaining issues affect the sustainment process.

Despite the challenges presented by the statutes, policies, and processes and the ever-evolving and never-settling engineering and data efforts described above, the senior members of the Air Force sustainment community understand their plight and are working to improve the sustainment enterprise. To this end, over the past several years, sustainment professionals have developed an overarching strategy called Expeditionary Logistics for the 21st Century (eLog21)—a compendium of governance, policies, processes and information that will be used for enterprise optimization. See Figure 2-1, which depicts the comprehensive nature of the eLog21 strategy. Its far-reaching goals will demand greater cooperation and integration

- AFI 63-101 - Acquisition and Sustainment Life-Cycle Management
  - Supersedes 12 Previous AFIs (10, 20, 21, 62, and 63 series)
  - 30 pages of Sustainment Planning Requirements
  - Life Cycle Systems Engineering
    - Includes Weapon Systems Integrity programs
    - Maintenance and Sustainment Engineering
- AFIs 20-101 and 63-1201 Tied to 63-101
  - In-depth Policies in Logistics Strategic Planning and Systems Engineering

FIGURE 2-7
Ongoing effort to consolidate the various policies. SOURCE: Colonel Kurt Hall, Deputy Director, Engineering and Technical Management Directorate, Headquarters Air Force Materiel Command, Wright-Patterson Air Force Base. "AFMC/EN Brief to the Air Force Studies Board." Presentation to the committee, December 8, 2010.

across the Air Force requirements, acquisition, engineering, and sustainment communities than in the past.

In addition, the Air Force's effort will have to fully permeate from the headquarters level to the execution levels. As shown in Figure 2-8, the eLog21 strategy is transformational and will require an extraordinary amount of procedural governance and discipline to achieve the envisioned integration of effort.

The details in Figure 2-9 clearly show that the each of the initiatives listed on the left has ample, detailed process flow information that will drive toward execution and completion. Other important initiatives have similarly detailed information charts.

With the ILCM instruction and the eLog21 initiative ongoing, Headquarters Air Force has established an Integrated Life Cycle Management Executive Forum, co-chaired by the Assistant Secretary of the Air Force for Acquisition (SAF/AQ) and the Assistant Secretary of the Air Force for Installations, Environment and Logistics (SAF/IE). Interestingly, depending on where a system is in its lifecycle, either SAF/AQ or SAF/IE has "51%" of the vote with regard to management and resource allocation decisions. This co-chair approach will go a long way to ensuring that

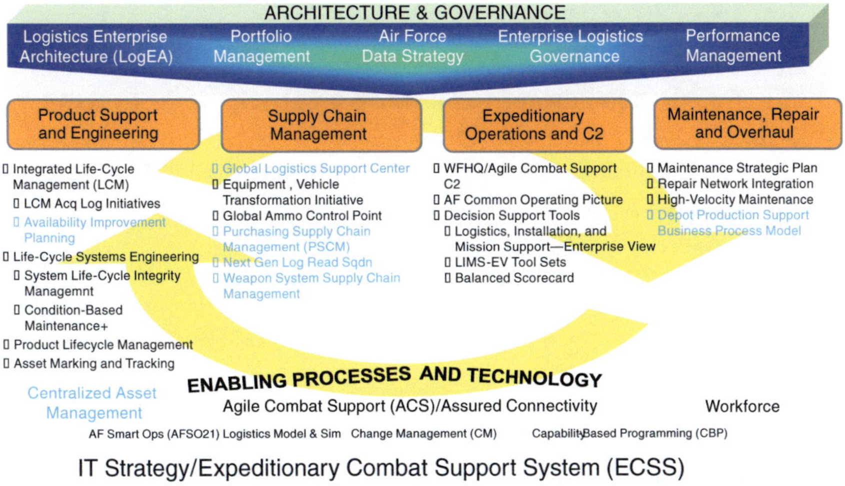

FIGURE 2-8
eLog21 Campaign plan. SOURCE: Grover L. Dunn, Director of Transformation, Deputy Chief of Staff for Logistics, Installations and Mission Support, Headquarters U.S. Air Force. "Expeditionary Logistics for the 21st Century (eLog21)." Presentation to the committee, January 17, 2011.

INITIATIVE	Focus
Establish Integrated Lifecycle Management Executive Forum	AF enterprise assessment and decision collaboration on life-cycle investment decisions to improve the integration of new products into the AF
Integrate AQ and LOG Policy and Enhanced Reporting	Combine and/or integrate AF acquisition and logistics policy to ensure seamless integrated life-cycle management. Ensure that life-cycle logistics assessment is consistently addressed in governance (e.g., SMART) and tools (e.g., POPs)
Improve Logistics Requirements and Traceability	Ensure that life-cycle logistics is addressed at every step from the lab to the requirements to the design and testing to the manufacturing and delivery process.
Acquisition and Logistics Architecture Integration	Integrate the Logistics Enterprise Architecture with the evolving Acquisition Enterprise Architecture to enable process and system integration and alignment throughout the product life cycle.

FIGURE 2-9
Product support initiatives. SOURCE: Grover L. Dunn, Director of Transformation, Deputy Chief of Staff for Logistics, Installations and Mission Support, Headquarters U.S. Air Force. "Expeditionary Logistics for the 21st Century (eLog21)." Presentation to the committee, January 17, 2011.

sustainment considerations are made a more important part of the development process for a new weapon system, which is clearly mandated in DoD Instruction 5000.02 and strongly recommended in the 2009 DoD *Weapon System Acquisition Reform Product Support Assessment* report.

Often during the course of the study, the opinion arose that the sustainment concept and strategy should be moved to much earlier in the acquisition process, that is, brought forward to the Milestone "A" Defense Acquisition Board decision. In fact, the DoD acquisition directive states the same, and the ILCM Executive Forum will help the Air Force ensure compliance with the directive. Furthermore, the *DoD Weapon System Acquisition Reform Product Support Assessment* and the LMI *Future Capability of DoD Depot Maintenance* reports comment on the state of acquisition and sustainment decisions and make recommendations for tighter controls and earlier sustainment considerations in the acquisition process, including making sustainment considerations a Milestone A exit criteria.[19,20] Under the eLog21 umbrella, the sustainment community works the Logistics Requirements Traceability initiative, which maps gaps and entry points to insert sustainment considerations into the early stages of the acquisition effort. This is another major step forward to ensuring that sustainment is considered as early as Milestone A in the acquisition process. The Executive Forum represents a significant advancement in governance, and the Air Force is to be commended for establishing this group. However, the committee found no factual evidence that the Air Force has a designated commander with responsibility for looking well beyond the year of execution to demand decision making that will leave the future Air Force in compliance with the earlier described statutes, regulations, and policies and with appropriately balanced and standardized long-term execution of the Air Force's sustainment directives. Clearly, the responsibility for developing Air Force policy rests with the Secretary of the Air Force. AFI 63-101 states that the Assistant Secretary of the Air Force for Installations, Environment and Logistics (SAF/IE) will:[21]

> Provide strategic logistics oversight for life cycle support; develop strategic level logistics, installations, and environmental policy for life cycle support; and provide vertical and horizontal integration of ILCM policies to provide for standardization and compliance

[19]DoD. 2009. DoD Weapon System Acquisition Reform Product Support Assessment. November. Washington, D.C.: Office of the Under Secretary of Defense for Acquisition, Technology, and Logistics. Available at https://dap.dau.mil/career/log/blogs/archive/2010/01/28/implementation-of-dod-weapon-system-acquisition-reform-product-support-assessment-psa-recommendations.aspx. Accessed November 22, 2010.

[20] LMI. 2011. Future Capability of DoD Maintenance Depots. LG901M2. February. Mc:ean, Virginia: LMI, p. v. Available at http://armedservices.house.gov/index.cfm/files/serve?File_id=394b31e6-4adc-47ca-a6f5-21547f0751fa. Accessed February 20, 2011.

[21]USAF. 2009. Acquisition and Sustainment Life Cycle Management. April 8. Available at http://www.e-publishing.af.mil/shared/media/epubs/AFI63-101.pdf. Accessed August 18, 2011.

> mechanisms across the Enterprise. Ensure functional policies as requested are communicated to the field. (p. 29)

Somewhat paradoxically, the same instruction states that the Headquarters Air Force/Directorate of Logistics will:

> Develop policy and issue AF implementation guidance for logistics support capabilities to ensure weapon system readiness for the user consistent with statutes, executive orders, and DOD issuances. Ensure functional policies as requested are communicated to the field. (p. 31)

The SAE authority for acquisition programs will not change, and the Assistant Secretary of the Air Force/Acquisition (SAF/AQ) is clearly the policy maker and executor for acquisition programs. AFI 63-101 also states that the SAF/AQ will

> Serve as the Service Acquisition Executive (SAE) as delegated for non-space AF programs and execute responsibilities as the senior corporate operating official for non-space acquisition. Execute SAE responsibilities outlined in the DOD 5000-series for execution of non-space AF acquisitions. For purposes of defining SAE responsibilities, this includes lifecycle acquisition of non-space systems and services processes from pre-Milestone A to weapon system retirement. This includes research, development, test, evaluation, production, and delivery of new systems, or significant modifications to existing systems. Management responsibility flows directly, without intervention, from the SAE and Milestone Decision Authority to the Program Executive Officers (PEOs) to the System Program Managers (SPMs). (p. 24)

Interestingly, these excerpts from the policy document do not use the terms sustainment or product support, but it is promising that they do use the term life cycle. Furthermore, the policy does not designate a single individual or commander as the sustainment or product support lead with responsibility for looking across the entire acquisition and sustainment enterprise (operational commands, in addition to the AFMC and the Defense Logistics Agency) and with authority for conducting future critical planning efforts, developing the execution details, standardizing engineering tools, and developing the sustainment equivalent of "operational tactics, techniques, and procedure." To ensure that AFI 63-101 is adhered to, the well thought out principles and processes of eLog21 are executed, and the significance of sustainment responsibilities are well represented, the Air Force should designate a field organization and commander as responsible for long-term sustainment planning and progression as well as for year of execution decision making. Without such a designation, the governance structure necessary to implement the lofty goals of eLog21 will drift, sustainment results in the out years will continue to be sub-optimized, and the Air Force will likely not achieve the eLog21 objectives, perceived sustainment goals, or the efficiencies mandated by fiscal year 2012 efficiency considerations.

Ironically, there is a need for some flexibility in the military environment. Therefore, as the committee discussed the issues associated with a senior sustainment commander, it recognized that there must be room for adjustments across the diverse operations. The committee believes that the sustainment enterprise would best work with strong centralized policy, procedures, and practices determined by the SAF/AQ, AF/A4/7, and the sustainment commander. When these are in place with the appropriate metrics, decentralized execution should be allowed to occur and performance reflected in the metrics.

Finding 2-5. The Air Force's eLog21 is an impressive umbrella effort encompassing multiple logistics transformation initiatives designed to help move the Air Force sustainment and acquisition communities toward an ILCM construct.

Finding 2-6. The SAF/IE, SAF/AQ, and AF/A4/7 effort to consolidate the key Air Force policy directives (AFPDs) dealing with acquisition and sustainment into a single AFPD is an important and positive step toward creating an ILCM culture in the Air Force. However, the Air Force has not yet completely developed the top-to-bottom governance structure, policies, and procedures necessary to achieve the ILCM objectives across the acquisition and sustainment enterprise.

Finding 2-7. It is not clear who has been designated as the sustainment process and product owner with authority to develop processes, advocate for enterprise sustainment interests, and assure balance in planning and execution of sustainment resources.

Recommendation 2-4. The Air Force should consider formally designating a senior commander, such as the AFMC, as the commander of the entire sustainment process, from concept phase through system retirement, with the responsibility to advise SAF/IE, SAF/AQ, and AF/A4/7 on policy and then train, organize, equip, plan, and execute the Air Force's ILCM processes.

Current Air Force Organization Complexities for Collaboration

Throughout the study, the question *"Who is in charge of sustainment?"* was repeatedly posed. As Figure 2-10 shows, there are numerous stakeholders in the Air Force sustainment enterprise. The organization is too complex for just a single program, let alone multiple programs. In fact, the organizational structure is so

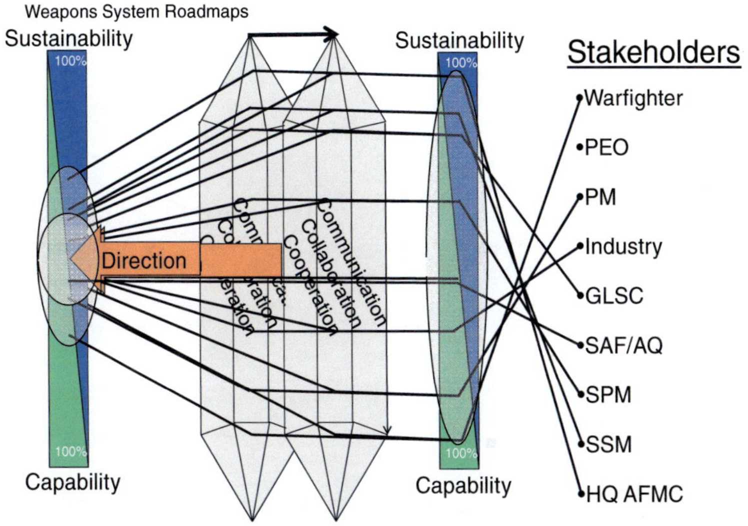

FIGURE 2-10
Focus the enterprise. GLSC, (Air Force) Global Logistics Support Center; HQ AFMC, Headquarters Air Force Materiel Command; PEO, Program Executive Office; PM, Program Manager; SAF/AQ, Assistant Secretary of the Air Force (Acquisition); SPM, System Program Manager; SSM, System Support Manager or System Sustainment Manager. SOURCE: Brigadier General Arnold W. Bunch, Jr., Director and Program Executive Officer for the Fighters and Bombers Directorate, Aeronautical Systems Center, Wright-Patterson Air Force Base. "ASC/WW (Fighter/Bomber PEO)." Presentation to the committee, December 8, 2011.

complex that collaboration, although essential, is unlikely to achieve repeatable successes. An accredited sustainment commander would be able to ensure appropriate streamlining and trans-organization coordination with significantly more effective results.

Outcome-based Metrics

At this stage in the chapter, five of the six criteria for an effective sustainment effort have been discussed. The logical remaining element is metrics for measuring and reporting on how the system and the processes are working. At the beginning of the chapter, the committee explained that the Air Force does not appear to have specified an overarching, outcome-based, measurable goal for the sustainment process. The committee did observe, however, the use of metrics across the sustainment enterprise. In fact, many very good metrics report on the various processes, and many focus on outcomes. Figures 2-11 and 2-12 illustrate the tracking of aircraft depot maintenance at Oklahoma City Air Logistics Center (OC-ALC). Figure

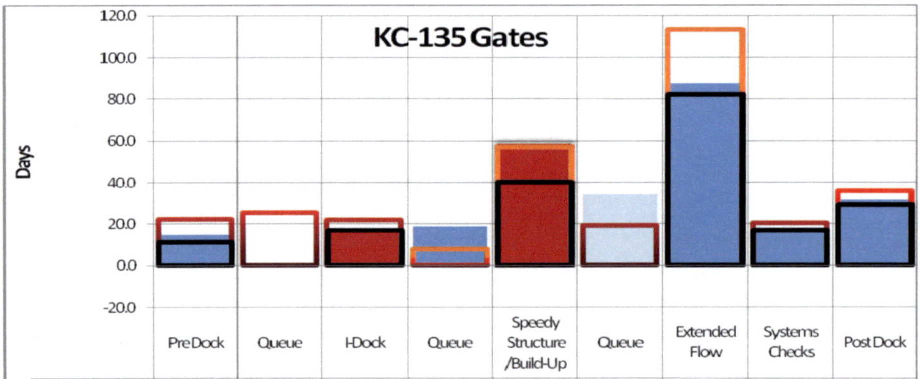

	PreDock	Queue	I-Dock	Queue	Speedy Structure /Build-Up	Queue	Extended Flow	Systems Checks	Post Dock
Last 10 A/C Avg	14.4	0.0	16.3	18.7	58.6	33.9	87.4	17.2	31.4
FY10 Qtr1	22.2	25.2	21.8	7.7	57.0	19.1	113.3	20.3	35.7
Requirement	11.0	0.0	17.0	0.0	40.0	0.0	82.0	17.0	29.0
%Change	-35%	-100%	-25%	143%	3%	77%	-23%	-15%	-12%

Overall Touch Time: -17% Overall Queue Time: +1%

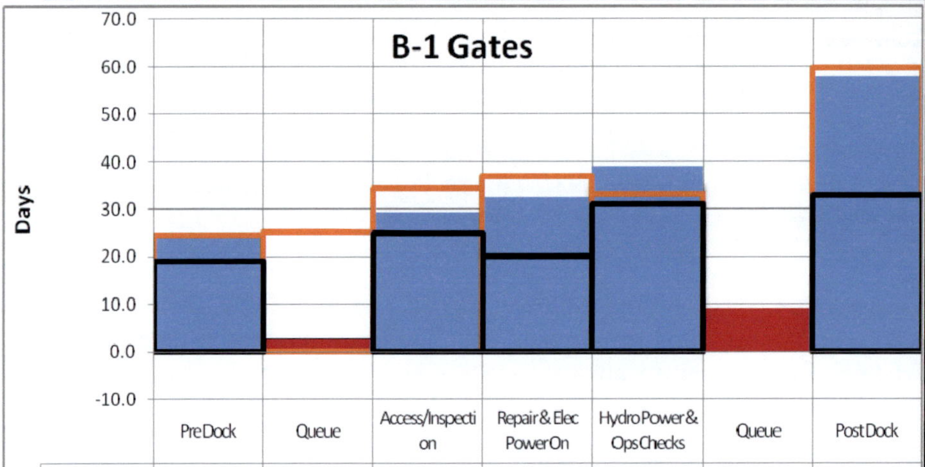

	PreDock	Queue	Access/Inspection	Repair & Elec PowerOn	HydroPower & OpsChecks	Queue	PostDock
Last 5 A/C Avg	25.0	2.6	29.4	32.4	39.0	9.2	58.0
FY10 Qtr1	24.5	25.2	34.5	37.0	33.2	0.0	59.6
Requirement	19.0	0.0	25.0	20.0	31.0	0.0	33.0
%Change	2%	-90%	-15%	-12%	18%		-3%

Overall Touch Time: -27% Overall Queue Time: -53%

FIGURE 2-11
The performance of a detailed top-level metric against an expected standard and an outcome-based expectation. SOURCE: Major General Bruce A. Litchfield, Commander, 76th Maintenance Wing, Oklahoma City Air Logistics Center, Tinker Air Force Base. "76 MXW Production Machine." Presentation to the committee, January 11, 2011.

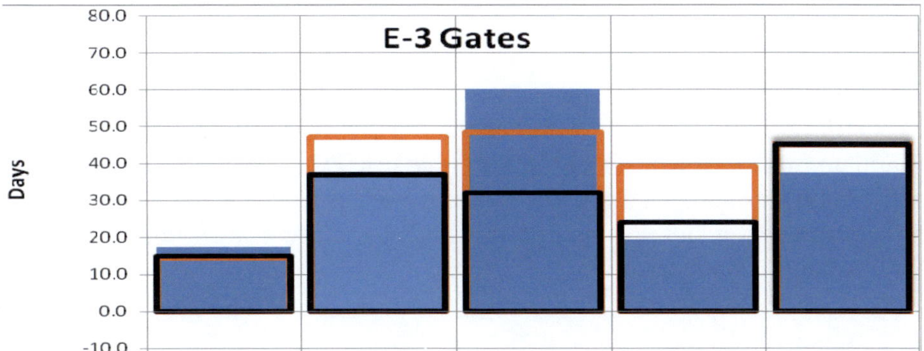

	PreDock	Inspection	Repair/ Build-Up	Final Ops	Post Dock
Last 3 A/C Avg	17.3	36.3	60.3	19.3	37.3
FY10 Qtr1	14.3	47.0	48.3	39.0	45.0
Requirement	15.0	37.0	32.0	24.0	45.0
%Change	21%	-23%	25%	51%	-17%

Overall Touch Time: -12% Overall Queue Time: N/A

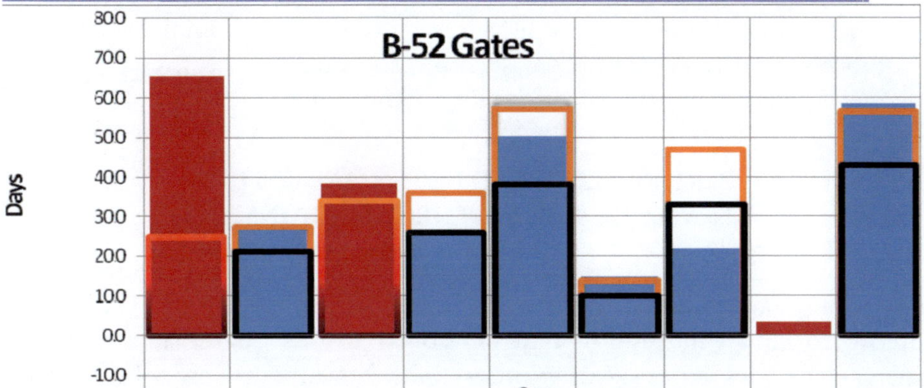

	Queue	PreDock	Queue	Inspection	Core Repair/ Restore	System Ops Check	SSR Work	Queue**	Post Dock
Last 6 A/C Avg	65.5	23.5	38.3	25.3	50.3	14.8	21.8	3.2	58.3
FY10 Qtr1	25.0	27.3	34.0	35.7	57.0	13.8	46.7	0.0	56.3
Requirement	0.0	21.0	0.0	26.0	38.0	10.0	33.0	0.0	43.0
%Change	162%	-14%	13%	29%	-12%	7%	-53%		4%

Overall Touch Time: -18% Overall Queue Time: +81%

FIGURE 2-11 *continued*

2-11 represents the performance of a detailed top-level metric against an expected standard and an outcome-based expectation. Figure 2-12 represents a "drill down" and analysis of the KC-135 product line covered in Figure 2-11.

As the committee studied these many metrics, it repeatedly asked if they build toward an overall assessment of the Air Force sustainment enterprise. In the end, the committee concluded that the metrics are well employed at the lower echelons of the organization, but they do not contribute to an overall assessment of the Air Force sustainment enterprise.

Finding 2-8. The Air Force sustainment enterprise employs extensive metrics to assess processes and performance, but these metrics do not build to an overall assessment of the Air Force sustainment enterprise.

Recommendation 2-5. The Air Force should develop and implement weapon system-level metrics that set AA levels and the cost of providing that availability, as well as identify who is responsible for attaining both. Furthermore, these measures should be at a level that reflects sustained implementation of process improvement initiatives, such as cost-reduction incentives and not just increasing sustainment costs driven by aircraft aging.

PAST, PRESENT, AND POTENTIAL FUTURE STRATEGIES FOR AIR FORCE SUSTAINMENT

The committee gave serious consideration to all aspects of the term "strategies" with respect to Air Force sustainment and determined that the strategies are interwoven and are the results—different from outcomes—of the statutes, regulations, and policies. From extensive interactions with Air Force officials, it became apparent that the strategies the Air Force either has in place, or desires to achieve, are designed to cope with the previously discussed statutes and policy constraints. For example, eLog21 is a strategy to shape processes and focus Air Force sustainment efforts on sustainment concepts and on producing a positive enterprise outcome (AA and cost) for the Air Force. Yet, the concept of how to gain an enterprise outcome for the direct support of weapon systems is a key part of the strategy equation. According to briefs from SAF/AQ, SAF/IE, AF/A4/7 and presentations from AFMC, enterprise weapon system support strategies are presently undefined and over the past 15 years have been program versus enterprise driven. These sustainment strategies range from total contractor support, to hybrid support from contractor and organic sources, to total government support. During these same discussions, it was noted how costly contractor support concepts are, but, at the same time, references were made to the positive outcome of the contractor support

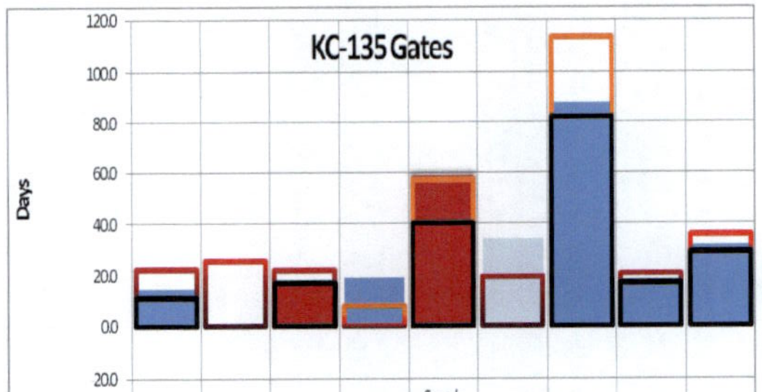

	PreDock	Queue	I-Dock	Queue	Speedy Structure /Build-Up	Queue	Extended Flow	Systems Checks	Post Dock
Last10A/CAvg	14.4	0.0	16.3	18.7	58.6	33.9	87.4	17.2	31.4
FY10Qtr1	22.2	25.2	21.8	7.7	57.0	19.1	113.3	20.3	35.7
Requirement	11.0	0.0	17.0	0.0	40.0	0.0	82.0	17.0	29.0
%Change	-35%	-100%	-25%	143%	3%	77%	-23%	-15%	-12%

Overall Touch Time: -17% Overall Queue Time: +1%

Top Constraints

•Constraint #1 Extended (SBUP) : Work content not resolved

•CPI #1 Extended (SBUP):

✓ Technical Resolution RIE to focus on Flow Day and Facility impact completed 17 May

✓Special Handling 202 process – 564 prototyping 202 process – added to Tinker Instruction ECD TBD

✓Daily stand-up meeting held for Andon aircraft

•Constraint #2 I-Dock: 80-100% parts not available in I-Dock

a.Landing Gear

•CPI #2 I-Dock:

✓MICAP, 202 Disposition for work-arounds

•Constraint #3 Speedy (SBUP) Queue: Late to need supply of critical parts

a.80-100% replacement parts Horizontal Hinge Shaft, Aft Quadrant

b.Lower % replacement parts AMARG request Fillet Assy

•CPI #3 Speedy (SBUP) Queue:

a.Visual Boards – C/W Prototype Board in Place Dock 2 – Remaining Boards in place 31 Oct 10 (POC: Theresa Farri

b.Replenishable Kits –

- Leading Edge/Trailing Edge Kit ECD 30 Oct 2010 (POC: Capt Ismirle)

FIGURE 2-12
"Drill down" and analysis of the KC-135 product line. SOURCE: Major General Bruce A. Litchfield, Commander, 76th Maintenance Wing, Oklahoma City Air Logistics Center, Tinker Air Force Base. "76 MXW Production Machine." Presentation to the committee, January 11, 2011.

concepts. Strategies were examined from the perspective of how the Air Force got to this state and what have been the true results.

As a part of its acquisition reform strategies in the early to mid-1990s, the DoD consciously selected operational systems to be entirely sustained (at least above the organizational level of support) by the commercial/contractor world. The Air Force already had been successful in sustaining some of its operational systems through Contractor Logistics Support (CLS), such as most of Air Education and Training Command's training aircraft, most of the operational support aircraft (of which there are at least 25 different types),[22] and the KC-10 tanker. Therefore, the Air Force initiated transitions of more systems to contactors, such as the C-17 under a Total System Performance Requirement (TSPR) contract followed by a CLS arrangement, or the mature F-117 system to a Total System Support Requirement (TSSR), which usually resulted in superior performance plus or minus the quality of the initial statement of work, defined metrics, and clear incentives. Missing from the equation was a balance between performance and cost: Although performance was generally considered excellent, the data were often owned by OEMs, and therefore future competitions were limited at best, and the cost factor appeared, to many individuals, to be expensive.

As these sustainment methodologies were designed and implemented, they fell neatly into line with the overall and emerging DoD preferred sustainment strategy known as Performance Based Logistics (PBL).

> The poster child of this latter approach (and by policy, DoD's preferred sustainment concept) is called Performance Based Logistics, more commonly referred to by its acronym PBL. PBL was and is transformative. Jacques Gansler, Under Secretary of Defense for Acquisition, Technology, and Logistics, from 1997 to 2001 described the context of what needed transforming at that time: ...to reverse this trend—with current short-term needs consuming an ever-increasing "share of the pie" at the expense of longer-term military capability—will be extremely difficult. I have called this situation a "death spiral;" and, in fact we will come to that...if we do not act decisively, now. It will require significant cultural change, a sense of urgency, and difficult program funding decisions. The result may be that we will have to put some sacred cows out to pasture—not just keep trying to milk them.[23]

It seems as though these findings are as true today as they were then.

As it turned over the sustainment of many of its systems to contractors, and in keeping with the general PBL philosophy, the Air Force moved to an outcome-

[22]Please see Tinker Air Force Base's Force Development Division, Oklahoma City Air Logistics Center website. Available at http://www.tinker-af.org/. Accessed May 4, 2011.

[23]DoD. 2009. DoD Weapon System Acquisition Reform Product Support Assessment. November. Washington, D.C.: Office of the Under Secretary of Defense for Acquisition, Technology and Logistics, p. 8. Available at https://dap.dau.mil/career/log/blogs/archive/2010/01/28/implementation-of-dod-weapon-system-acquisition-reform-product-support-assessment-psa-recommendations.aspx. Accessed November 22, 2010.

based concept. In doing so, the Air Force did not always stipulate the needs for individual contractors to use data gathering and analysis systems that would be compatible with Air Force systems, for specific engineering processes to continually review the fleet status, and for altering maintenance and sustainment processes, as well as the Technical Orders. The Air Force did not always have a contract for access to second- and third-level of engineering and manufacturing data and the details necessary to understand the materiel management and/or maintenance management status with regard to contractor-sustained operational systems. Having the proper data is essential to determining a contractor and Air Force blended work-sharing approach.

As time went on, and as the Air Force altered the statements of work to fine tune/reduce the costs of operating the systems, it found re-competitions to be very challenging—if not impossible—because it did not have command of the data used by the incumbent (often OEMs) to sustain the systems. This effort to transition to different contractors or to bring maintenance efforts under organic support has suffered several fits and starts during the past 10 years. As shown in Figure 2-13, the Air Force is reaching the conclusion that contractor support arrangements are prohibitively expensive.[24] This is one perspective, but the committee has observed a different perspective.

Although the resources did not exist to extensively examine the cost versus effectiveness trade-offs, the available data indicated that the true challenge is in ensuring that comparisons of effectiveness and efficiency are objective. For example, the data presented in Figure 2-14, adjusted as noted, indicate that compared to other organically supported and contractor-supported platforms, the C-17 offers tremendous performance.[25]

Figure 2-15 depicts the performance of the C-17 in terms of reductions in cost per flying hour. These reductions were due to several factors, including more aircraft and more flying hours to cover fixed costs and changes in contract structure and performance standards. Although these data elements are important and accurate, a true measure of performance also must account for the platform's effectiveness. Figure 2-16 illustrates the demand on this high-value aircraft system as well as all Air Mobility Command aircraft.

The C-17 is the workhorse of the air mobility fleet and is in constant demand. Of course, the Air Force has finite resources, but concerns about sustainment costs must be weighed against what is being delivered in terms of capability. Further-

[24]Deborah K. Tune, Principal Deputy Assistant Secretary of the Air Force for Installations, Environment and Logistics, Office of the Assistant Secretary of the Air Force for Installations, Environment and Logistics. "Air Force Studies Board Sustainment Study: Developing the Right Product Support Concepts for the Future." Presentation to the committee, October 10, 2011.

[25]The "assumptions" for the costs associated with Figure 2-15 are much different from the assumptions made with Figure 3-6.

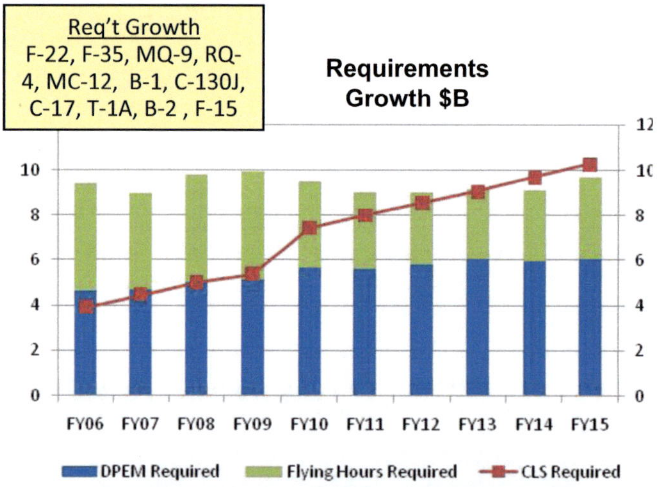

TAI			
	FY00	FY15	Growth
DPEM	5477	4491	-18%
CLS	729	1669	129%

DPEM vs. CLS TAI growth

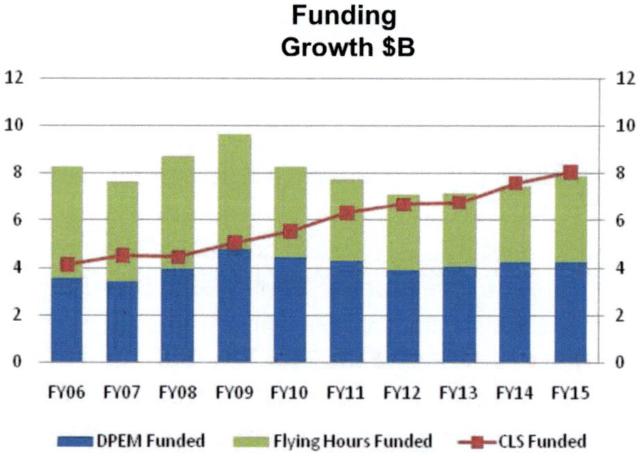

CLS Top Funding Drivers
F-22, F-35, ISR, W/S MIPs

FIGURE 2-13
Organic versus contract sustainment. SOURCE: Debra K. Tune, Principal Deputy Assistant Secretary of the Air Force for Installations, Environment and Logistics, Office of the Assistant Secretary of the Air Force for Installations, Environment and Logistics. "Developing the Right Product Support Concepts for the Future." Presentation to the committee, October 20, 2010.

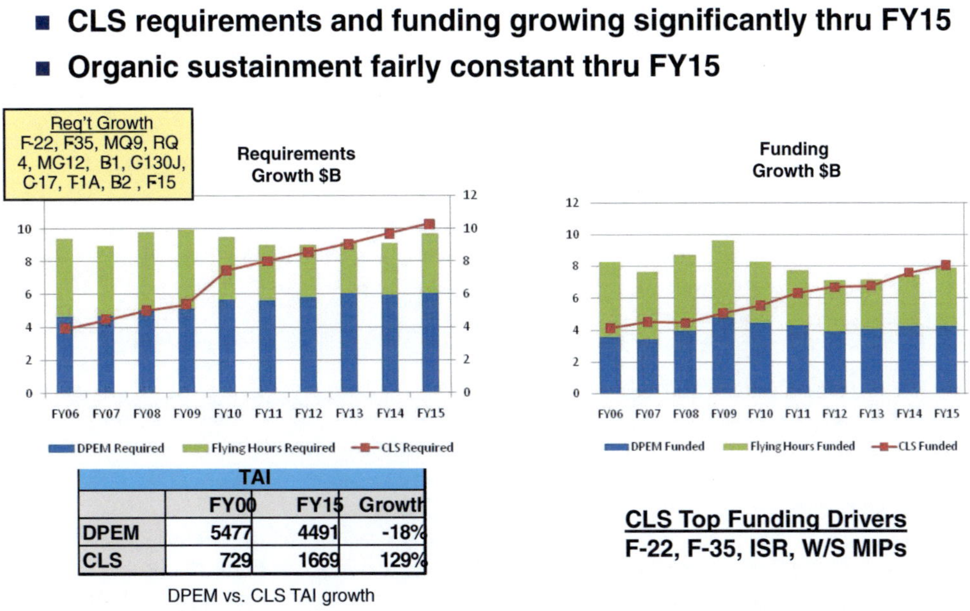

FIGURE 2-14
Change in weapon system cost from FY2004 to FY2009. SOURCE: Mark Angelo, Director, C-17 GSP Operations and Site Lead, The Boeing Company; Robert Tomilowitz, Executive Director, Supply Chain Management and Support Equipment, The Boeing Company; and Richard (Skip) Whittington, Senior Manager, C-17 GSP Business Development, Boeing Defense, Space & Security. "C-17 Globemaster III Sustainment Partnership." Presentation to the committee, January 18, 2011. Personal communication with the committee by the C-17 program office personnel who confirmed the accuracy of the data. Personal communication by the committee with Gustavo Urzua, The Boeing Company, who acknowledged that the data source was the Air Force total ownership cost database.

more, the 2009 *DoD Weapon System Acquisition Reform Product Support Assessment* analyzed partnering strategies and stated:

> While there are a range of indicators resulting from the maturity assessments and root-cause analysis, the weapon system data analysis clearly shows that performance-based (outcome-based) product support strategies, particularly when coupled with government-industry partnering approaches, have consistently delivered improved materiel readiness across numerous weapon system applications over the past decade. Cost benefits are more difficult to assess; as cited in several GAO [Government Accountability Office] reports, many outcome-based support strategies have claimed cost reductions and cost avoidance, but DoD financial systems lack the visibility and fidelity to validate these benefits consistent with audit standards. In summary, performance-based product support strategies consis-

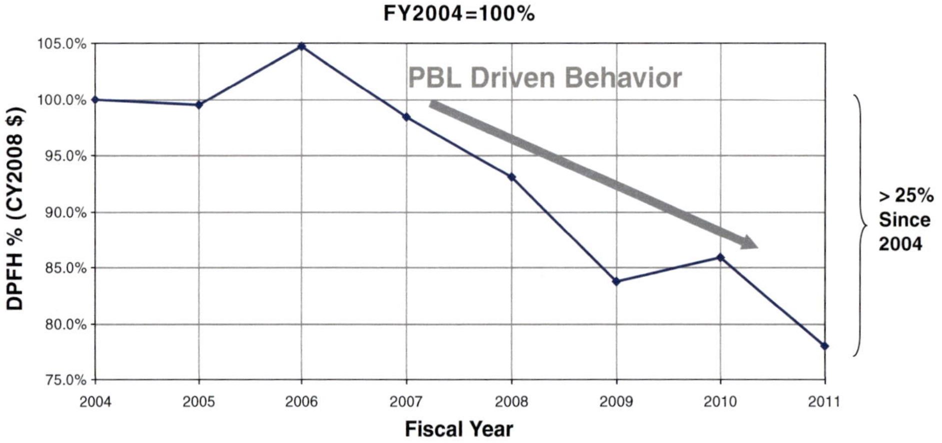

FIGURE 2-15
C-17 Global Sustainment Partnership (GSP): Dollars per flight hour (DPFH) reductions FY2004-FY2011. SOURCE: Mark Angelo, Director, C-17 GSP Operations & Site Lead, The Boeing Company; Robert Tomilowitz, Executive Director, Supply Chain Management and Support Equipment, The Boeing Company; and Richard (Skip) Whittington, Senior Manager, C-17 GSP Business Development, Boeing Defense, Space & Security. "C-17 Globemaster III Sustainment Partnership." Presentation to the committee, January 18, 2011.

FIGURE 2-16
Demand on mobility fleet FY2002-FY2009. SOURCE: Timothy Thomas, Deputy Chief, Maintenance Division, Headquarters Air Mobility Command. "AMC/A4 Perspective." Presentation to the committee, October 21, 2010.

> tently deliver improved materiel readiness, but assessing the true cost of both traditional (transactional) and performance-based strategies is difficult, if not impossible, given current financial systems.[26]

Additionally, as Figure 2-17 shows, strong performance factors are derived from partnering relationships. Cost is always a factor, but the partnerships must capitalize on each of the partners' strengths and deliver an outcome to the warfighter that is both highly effective and efficient. In striving to deliver effectiveness and efficiency, the DoD and Air Force must consider external factors that affect sustainment strategies, largely the legislative provisions of Title 10 of the U.S. Code. At the same time, 10 U.S.C. § 2474 offers significant relief by allowing the Air Force to establish partnering relationships between maintenance depots and contractors in the form of Centers of Industrial and Technical Excellence. Although this provision has existed since November 1997,[27] only 1 percent of the Air Force's depot maintenance work is conducted by public-private partnerships.[28] The intent of the Congress in passing this legislation was to allow Centers of Industrial Technologies and Excellence—Air Force depot maintenance facilities, in this case—to partner with industry and share work in the government facilities where both could gain work and the government facilities could gain technology enhancements to allow them to take on future work. The workload is taken out of the 50/50 calculations.

[26]DoD. 2009. DoD Weapon System Acquisition Reform Product Support Assessment. November. Washington, D.C.: Office of the Under Secretary of Defense for Acquisition, Technology and Logistics, p. 11. Available at https://dap.dau.mil/career/log/blogs/archive/2010/01/28/implementation-of-dod-weapon-system-acquisition-reform-product-support-assessment-psa-recommendations.aspx. Accessed November 22, 2010.

[27]Quoting, in part, 10 USC § 2474 entitled "Centers of Industrial and Technical Excellence: Designation: Public-Private Partnerships" and enacted November 18, 1997, by Public Law 105-58: "(1) The Secretary concerned, or the Secretary of Defense in the case of a Defense Agency, shall designate each depot-level activity of the military departments and the Defense Agencies (other than facilities approved for closure or major realignment under the Defense Base Closure and Realignment Act of 1990 (part A of title XXIX of Public Law 101-510; 10 U.S.C. 2687 note)) as a Center of Industrial and Technical Excellence in the recognized core competencies of the designee. (2) The Secretary of Defense shall establish a policy to encourage the Secretary of each military department and the head of each Defense Agency to reengineer industrial processes and adopt best-business practices at their Centers of Industrial and Technical Excellence in connection with their core competency requirements, so as to serve as recognized leaders in their core competencies throughout the Department of Defense and in the national technology and industrial base (as defined in section 2500(1) of this title [10 USCS § 2500(1)]). For additional information, see http://www.law.cornell.edu/uscode/html/uscode10/usc_sec_10_00002474----000-.html. Accessed July 19, 2011.

[28]LMI. 2011. Future Capability of DoD Maintenance Depots. LG901M2. February. McLean, Virginia: LMI, pp. 4-15. Available at http://armedservices.house.gov/index.cfm/files/serve?File_id=394b31e6-4adc-47ca-a6f5-21547f0751fa. Accessed February 20, 2011.

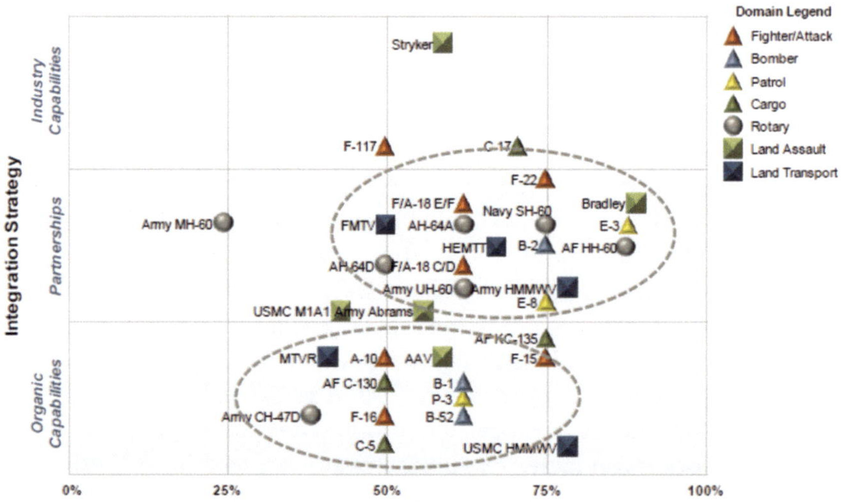

FIGURE 2-17
Partnering strategies produce higher sustained readiness improvement. Notes: 1. Sustained Readiness Improvement is the number of years over the span of 1999 through 2007 where a weapon system saw no decline in availability or saw a decline of lesser magnitude than the domain average. 2. F-22, FMTV, MTVR, and Stryker data do not span from 1999 through 2007 due to their newness. 3. USAF C-130 APU contract awarded to Honeywell in August 2007. Not enough time has occurred yet to include it as a partnership for this evaluation. SOURCE: DoD. 2009. DoD Weapon System Acquisition Reform Product Support Assessment. November. Washington, D.C.: Office of the Under Secretary of Defense for Acquisition, Technology and Logistics. Available at https://dap.dau.mil/career/log/blogs/archive/2010/01/28/implementation-of-dod-weapon-system-acquisition-reform-product-support-assessment-psa-recommendations.aspx. Accessed November 22, 2010.

Finding 2-9. The Air Force has used a variety of sustainment practices for new weapon systems and major modifications on older weapon systems without due consideration of their impacts on the sustainment enterprise. This has resulted in a wide diversity of sustainment practices and difficulty in tracking and understanding true sustainment costs.

Recommendation 2-6. The Air Force should develop and direct a sustainment execution model for weapon systems and major weapon systems modifications that balances the needs of the individual weapon system with the performance goals and cost constraints of the overall sustainment enterprise.

A BENCHMARK: PROCESSES AND POLICIES OF THE NAVAL AVIATION ENTERPRISE

There are, indeed, wide differences in organization, size, and culture between the United States Navy (USN) and the Air Force aviation communities.[29] The committee elected to examine how the Navy conducts sustainment operations. The committee did so briefly, not knowing what it would find, but recognizing that benchmarking—good or bad—affects thought processes. The Navy faced challenges in the late 1990s and early 2000s much like those faced by the Air Force today—in terms of readiness, support, and understanding and controlling costs. As a result, the Navy initiated a complete review of, and revision to, its approach to aviation readiness and sustainment. Simply put, the Navy focused on strategic outcomes—readiness and cost—or, in other words, "Be ready, control consumption."

This review was necessitated by the observation of the Chief of Naval Operations that principal performance indicators for the aviation community were consumption oriented and had no value in measuring readiness.[30] In addition, only limited performance indicators existed to provide Navy leadership with information about the cost-effectiveness of existing actions to provide readiness, establish outcome-based budget requirements, and manage operational results. In essence, readiness was not adequately, or appropriately, defined or measured in terms of its contribution, performance, and cost, relative to the Navy's goals.[31] As a result of the review, Naval Aviation was aligned into the Naval Aviation Enterprise (NAE). The NAE represents a behavior model focused on the warfighter as the Single Process Owner (SPO), with Navy resource and provider organizations in subordinate roles.[32] Figures 2-18 and 2-19 provide graphic portrayals of the NAE.

At the apex of the pyramid is the operational commander or supported commander who sets the tempo for the overall operation and is in fact the NAE commander. The many providers who support the commander are on the left-hand leg. The resource sponsors, who provide the resources to conduct operations, are on the right-hand leg. Interestingly, although the acquisition arm is under the Naval Air Systems command and is a force supporter, the maintenance depots and

[29]Appendix C, Navy Enterprise Transformation, provides additional details on the naval Aviation Enterprise.

[30]USN. 2007. Navy Enterprise Transformation: Working for the Greater Good, pp. 3-6. May. Available at http://www.thomasgroup.com/getdoc/7c79c3c9-8603-4908-ad89-0ebb0e10a67f/Navy-Enterprise-Transformation.aspx.

[31]Ibid.

[32]A similar positive relationship is discussed in Chapter 6 on page 6-29 where the triad of the FAA + OEM + Operator is established in the commercial aviation sector to achieve what the Navy implemented in the NAE. More specifically, the arrangement brings together the key elements of operating a fleet, including sustainment, to achieve an enterprise solution.

❖ **Enterprise Culture and Communication**

- Achieve a culture that emphasizes and rewards collaboration, accountability and transparency among Enterprise stakeholders and partners in support of Naval Aviation readiness. (Champions: CNAF, DC (A), NAVAIR)

❖ **Readiness**

- Engage all Naval Aviation readiness stakeholders and stakeholder organizations to drive efficient delivery of combat-ready forces to meet current and future operational requirements. (Champions: CR CFT Co-Leads)

❖ **People**

- Develop and manage enterprise-wide processes that maintain a diverse, cost-effective, and technically superior total force to perform all of the functions required for Naval Aviation to fight and win in combat. (Champion: CNATRA)

❖ **Future Readiness**

- Engage stakeholders to effectively produce required levels of future readiness while optimizing costs. (Champion: OPNAV N88)

FIGURE 2-18
Strategic objectives of the Naval Aviation Enterprise (NAE). SOURCE: Captain Mike Kelly, Commander, Naval Air Forces (CNAF), Force Material, Maintenance and Readiness. "COMNAVAIR." Presentation to the committee, January 17, 2011.

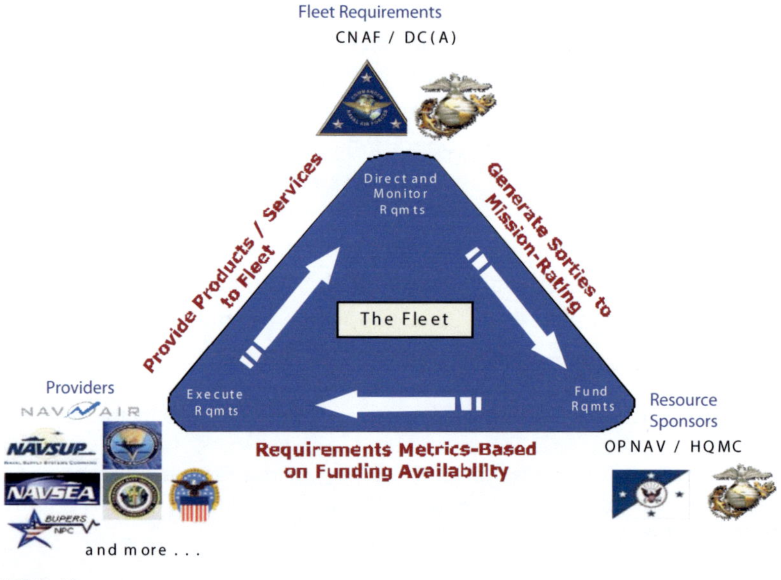

FIGURE 2-19
NAE's enterprise framework. SOURCE: Captain Mike Kelly, Commander, Naval Air Forces (CNAF), Force Material, Maintenance and Readiness. "COMNAVAIR." Presentation to the committee, January 17, 2011.

operational-level maintainers directly report to the commander of the Naval Air Forces even though they are providers.[33]

The NAE's focus shifted from consumption (accumulation of flight time, carrier landings, and experience) to metrics (Single Fleet Driven Metric, or SFDM), reflecting the state of the fleet relative to goals established collectively by the fleet and senior military and government leadership. The organizational alignment and commitment to common metrics served to eliminate the "stovepipe" orientation common to many functionally driven organizations and focused all Naval Aviation assets on a holistic approach to readiness as shown in Figure 2-20.

As an example, a critical element involved a change to the relationship between the fleet, i.e., the operators, and the maintenance providers. Wing commodores became the principals for their aircraft types and were expected to assume responsibility for readiness, with maintenance providers subordinated to that effort.

Demonstratively representative statements of the NAE's objective and organizational and philosophical approach are found in the minutes of the executive committee's May 2010 meeting:

> "Every time we do this meeting, we find that there is continuing relevance and there is a continuing reason for the Naval Aviation Enterprise," said Vice Adm. Thomas J. Kilcline, Jr., commander, Naval Air Forces. "The NAE is maturing, and every year over the last six years, we have taken a really good hard look at what we are doing and how we can improve. This year, as in the past, we questioned the assumptions that underline what we do and why we do it."
>
> "The NAE is relevant, if not more relevant today, in championing the processes that deliver a warfighting capability. The purpose of this meeting is to bring together a core group of leaders that have an impact on naval aviation," said James Beebe, executive director for Commander, Naval Air Forces. "It is all about communication and understanding the equities that we all have in supporting naval aviation requirements."[34]

The NAE is to provide a leadership/management alignment that assures all functions related to the warfighting capability of Naval Aviation are focused on the warfighter's missions (as demonstrated in Figure 2-21). The concept includes corporate tools and methodologies to integrate financial considerations and measurements into the operational process as a means to address fiscal and budgeting realities. A plan approved by a Board of Directors (BOD), consisting of the most senior leaders of the functional activities, exists for each functional alignment and identifies, prioritizes, aligns, and synchronizes investment efforts for the various entities. Gaps in capabilities relative to need are addressed with action plans. Prior-

---

[33]USN. 2007. Navy Enterprise Transformation: Working for the Greater Good, pp. 3-6. May. Available at http://www.thomasgroup.com/getdoc/7c79c3c9-8603-4908-ad89-0ebb0e10a67f/Navy-Enterprise-Transformation.aspx.

[34]Available at http://www.navy.mil/search/display.asp?story_id=53356. Accessed May 1, 2011.

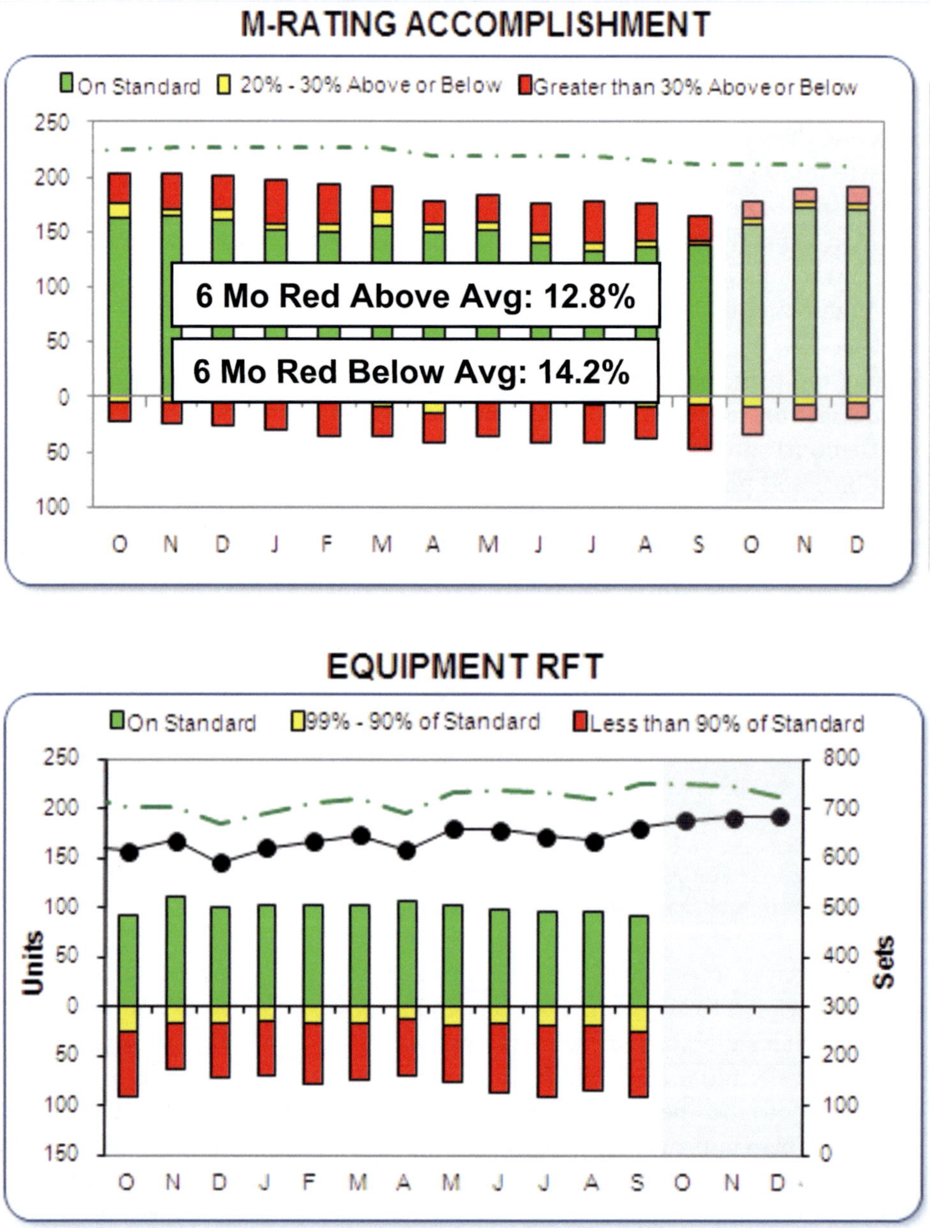

FIGURE 2-20
Current readiness Big 4. SOURCE: Captain Mike Kelly, Commander, Naval Air Forces (CNAF), Force Material, Maintenance and Readiness. "COMNAVAIR." Presentation to the committee, January 17, 2011.

Enlisted Personnel RFT

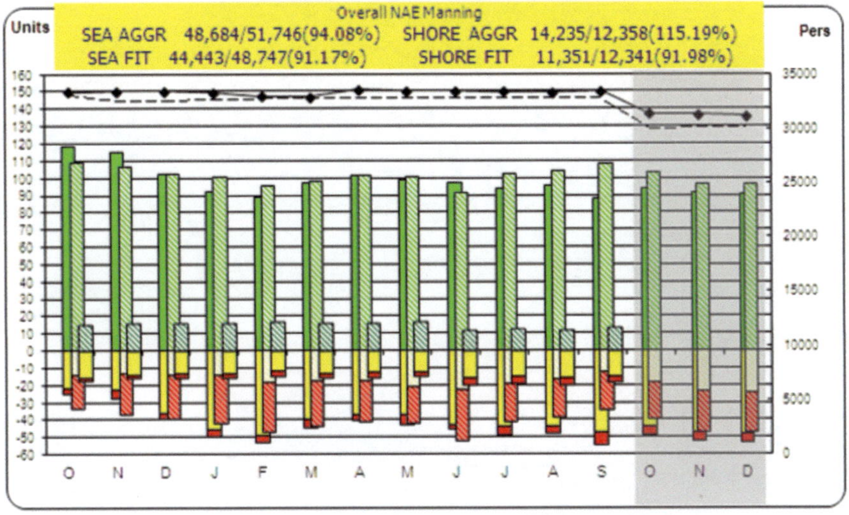

COST PERFORMANCE

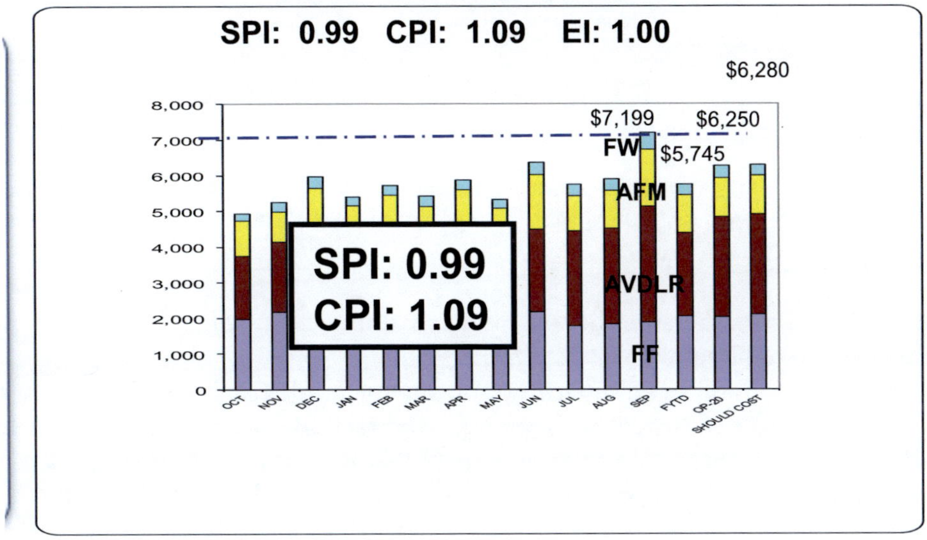

FIGURE 2-20 *continued*

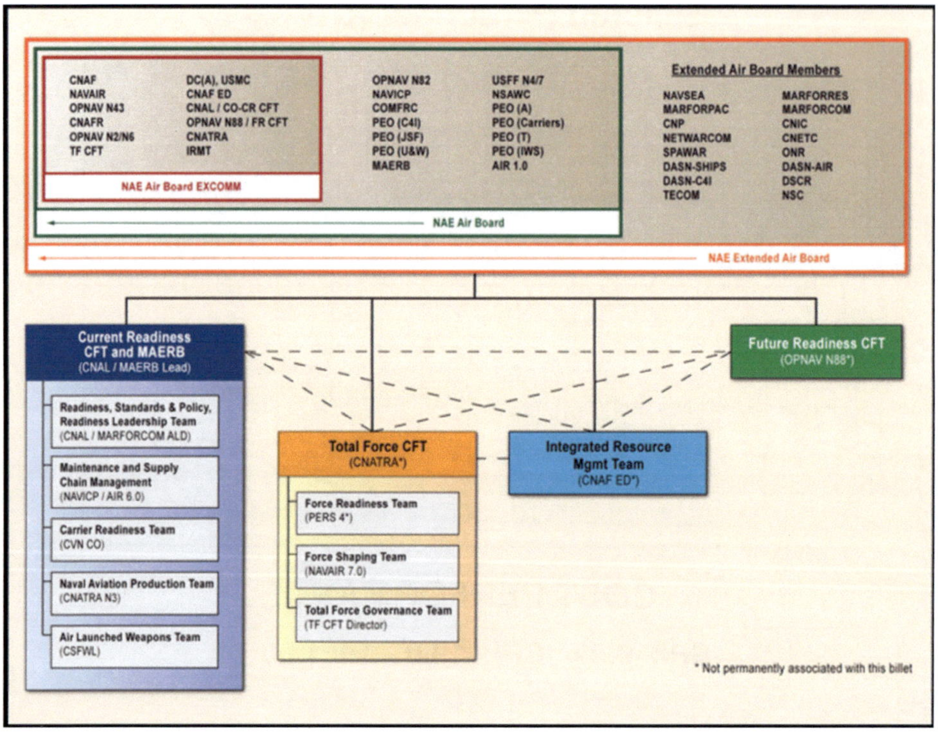

FIGURE 2-21
Today's NAE. SOURCE: Captain Mike Kelly, Commander, Naval Air Forces (CNAF), Force Material, Maintenance and Readiness. "COMNAVAIR." Presentation to the committee, January 17, 2011.

ity and resource differences between functional units that cannot be resolved by the NAE are resolved by the BOD for the NAE.

Readiness is addressed in terms of operational goals established by the most senior leadership consistent with strategic and tactical doctrine. Sustainment is addressed through the NAE triad alignment that focuses the efforts of the Technology Developer (ONR/NRL/NRE) and the Material Developer (NAVAIR) on the needs of the Combat Developer (CNAF/OPNAV/CFFC) to optimize maintenance processes and support equipment and to provide a holistic approach to meeting required operational standards. The same model aligns personnel recruitment and training with operational needs. The focus is on meeting established operational objectives, or exceeding those objectives within fixed cost budgets, and management constraints are exerted to prevent doing more than is necessary to surpass those goals for which variable costs are involved. Evidence of the process' success includes the ability of the NAE to meet operational objectives with a surplus of manpower, financial, material, flying, and other resources that have been eliminated or returned for disposition to senior leadership. The NAE established the behavior model, organization, and processes to[35]

1. Provide clear definition, organization, and planning process to achieve the Single Fleet Driven Metrics for readiness and sustainment.
2. Provide a holistic methodology for organizational and functional alignment under appropriate warfighter commands, as Single Process Owners, to focus all asset efforts on the objectives established by the NAE.
3. Provide fiscal/budgetary awareness relative to operational objectives within all aviation commands to aid in preventing unnecessary variable cost expenditures for unneeded readiness or sustainment capability and provide improved readiness at reduced cost.
4. Provide a methodology for governance and continuous planning to address action plans to achieve current and future operational readiness, sustainment, and fiscal objectives in a process of continuous improvement.
5. Provide SFDM as Key Performance Indicators for transparency within the entire Naval Aviation community relative to readiness performance.

Finding 2-10. Although the Air Force structure and program management mechanisms are designed differently, the NAE approach provides an interesting governance model and foci for the Air Force to consider as it executes its eLog21 strategy and deliberates on how to sustain its overall force.

CONCLUDING THOUGHTS

Foremost, the lack of clearly defined sustainment goals affects the entire Air Force. The Air Force can quickly develop a solution to this situation and then refine it over the longer term. The Air Force would be well served by assigning a sustainment commander with stature to work with the operational commands as well as the SAF/AQ, SAF/IE, and AF/A4/7 to develop and vet policy. This commander could then dynamically shape long-term sustainment planning, be accountable for achieving the sustainment goals, and oversee execution of the Title 10 U.S.C. provisions and day-to-day support of the fleet. From the outset of the study, the committee compared the Air Force sustainment model to general operations of large commercial firms. As stated earlier, it is imperative to determine who is in charge and what are the roles and consequences of their being in charge. Figure 2-22 notionally provides a side-by-side comparison of commercial business operations and accountability to what is considered to be the Air Force model.

[35]Vice Admiral Walter Massenburg (United States Navy, retired), Senior Director, Mission Assurance Business Execution, Raytheon Company. "Enterprise Behavior—Fundamental Changes in the Government Business Model." Presentation presented to the committee, January 18, 2011.

	COMMERCIAL	USAF
GOAL	Strategic vision Maximize profits Solid ROI	Not defined or unclear Optimized combat status (C-Rating
METRIC	Costs and profitability Operational performance Customer satisfaction	Varies Aircraft availability (AA), Mission-capable rates
RESPONSIBLE	CEO	Too many to list
JUDGE	Wall Street	SECAF/CSAF
CONSEQUENCE	Reward or Replacement	None unless total failure occurs e.g., Nuclear Enterprise Breakdown

FIGURE 2-22
Notional model depicting Air Force and commercial approaches to aircraft sustainment.

Any system in an organization the size of the Air Force is extraordinarily complex. Based on its interactions with numerous Air Force officials, the committee is confident that the officials understand what they are doing and the importance of their duties to the nation. At the same time, they expressed a general frustration with an inability to control their work processes and achieve positive results.

Challenges in the Air Force sustainment process begin in the SAF/AQ, SAF/IE, and AF/A4/7 offices, which are responsible for defining policy and procedures for the subordinate organizations. These headquarters offices must set the tone for Air Force sustainment. In the absence of well-defined policy and procedures, field-level commanders and directors take individual action to sustain their fleet. Such initiative should be noticed, but the commanders and directors should receive clear guidance and should be held accountable for execution. Statutes have sometimes served as an excuse from making hard decisions, and these decisions do involve economic trade-offs. Nevertheless the statutes will remain, and the Air Force must set the stage for long-term compliance.

The Air Force sustainment process is not broken by statute, policies, regulations, or strategies. However, it is not an efficient enterprise either. As reflected in several recommendations, the sustainment process can be improved, but to do so will require continued efforts through programs such as eLog 21 and SAF/AQ initiatives. Yet, these alone will not address all issues or even solve the key issue. For the enterprise to truly function as an enterprise, it needs a strong leader…a leader that is in charge of *both* planning and execution.

3

Assessment of Current Sustainment Investments, Infrastructure, and Processes

INTRODUCTION

This chapter addresses element 1 of the terms of reference (TOR), that is, "Assess current sustainment investments, infrastructure, and processes for adequacy in sustaining aging legacy systems and their support equipment." This chapter also sets the stage for other chapters about the importance of the funding that significantly impacts the sustainment of systems over the entire life cycle. Sustainment investments are found in multiple elements of the Air Force budget in part because of the fact that sustainment cuts across all aspects of weapon system life cycles.

The Air Force faces a number of sustainment challenges, including aging of aircraft systems, rapid advances in technology, and increasing costs. As noted in Chapters 1 and 2, the Air Force does not have a clearly articulated definition of sustainment or high-level sustainment goals. Instead, most leaders involved in sustainment-related activities define their goals only in terms of achieving an availability metric at the platform level. There was not a clear understanding of the source of the availability requirement for each platform or of the relationship between aircraft availability and cost by platform or across the entire enterprise.

AIR FORCE INVESTMENT PROCESS RELATING TO SUSTAINMENT

Department of Defense (DoD) and Air Force budgeting and funding are institutionalized processes that respond to national needs in terms of strategies and doctrines and include the functional activities of operation and maintenance

(O&M), procurement, research, development, testing, and evaluation (RDT&E), and military construction that have either a direct or indirect relationship to sustainment. The O&M portion of the budget is usually associated with sustainment because it is within this realm that the most visible activities associated with the metric of aircraft availability are funded. It also is in this budget area that investments in depot maintenance, some hardware procurement, logistics, and, in some cases, contract logistics support are made.

The procurement phase of weapon system programs has historically been driven by cost and performance parameters that are established in the concept, or pre-Milestone A, phase in Defense Acquisition Board (DAB) decisions, when program budgets are established on a cost-benefit basis. Only recently, as exemplified by the Air Force memorandum "Present a Competitive Acquisition Strategy at Each Program Milestone," has the Air Force stressed the importance of incorporating sustainment considerations into all phases of the acquisition process.[1] The importance of addressing sustainment early in the acquisition process cannot be over-emphasized and was a subject raised repeatedly during the study. Consider the following quotes from the 2009 *DoD Weapon System Acquisition Reform Product Support Assessment*:

> Acquisition processes pay too little attention to supportability and consistently trade down-stream sustainability for required capability or program survival. Some Program Managers assert that "logistics is their only discretionary account", making it a frequent target for inevitable resource reductions. In acquisition decision reviews, sustainment is often relegated to the back-up charts. Hampered by functionally stove-piped organizational structures and lacking life cycle management qualifications in their diverse workforce, the logistics community fails to achieve effectively integrated and affordable Warfighter operational readiness. Instead, it remains focused on managing commodities, parts, and services.[2] (p. 7)

> Product support, vital to both acquisition and logistics, has been treated as the stepchild of both functions. The acquisition community has neglected it, and the logistics community seems mismatched to effectively perform its demanding scope.[3] (p. 7)

---

[1]DoD. "Present a Competitive Acquisition Strategy at Each Program Milestone." A Memorandum from Air Force Service Acquisition Executive David M. Van Buren to Senior Air Force Personnel. January 14, 2011. Washington, D.C.: Office of the Assistant Secretary. Available at https://acc.dau.mil/CommunityBrowser.aspx?id=433566. Accessed March 23, 2011.

[2]DoD. 2009. DoD Weapon System Acquisition Reform Product Support Assessment. Washington, D.C.: Office of the Secretary of Defense. November, p. 7. Available at https://acc.dau.mil/adl/en-US/328610/file/47489/DoD%20Weapon%20System%20Acquisition%20Reform%20PSA_19%20NOV_Final.pdf. Accessed March 22, 2011.

[3]Ibid.

The Air Force is attempting to put in place new acquisition processes—for example, giving the logistics community a seat at the table earlier in the acquisition process.[4] The Air Force Logistics Requirements Traceability (LRT) process is intended to "ensure that life cycle logistics is addressed at every step from the lab to the requirements to the design and testing to the manufacturing and delivery process."[5] The white circles in Figure 3-1 show activities leading up to Milestone A, where there is currently inadequate logistics representation or assessment. As part of the LRT process, a set of standard work, tools, and templates is being developed to ensure that logistics requirements are addressed and tracked over the complete lifecycle.

As noted in a recent article, ". . . while attention is typically focused on the initial cost of procuring a weapon system, sustainment spending actually accounts for most of the total lifetime cost of ownership. Sustainment is critical from a mission and readiness perspective. When sustainment is optimized, weapon systems perform better, spend less time under repair, and remain in use longer, thus delaying the need for their replacement." [6]

From a life-cycle cost perspective, early consideration of sustainment as part of a comprehensive approach to systems engineering is critical. As shown in Figure 3-2, only a small fraction of the overall life-cycle budget is spent during the concept development phase of a program; however, by the time that phase is completed approximately 70 percent of the life-cycle cost of the program is committed. Based on a statistical analysis performed on DoD projects, the Defense Acquisition University reports that by the time 20 percent of the life-cycle cost has been spent more than 80 percent of the life-cycle cost has been committed. Early involvement by the sustainment community offers the maximum opportunity to leverage its insights and experience to impact life-cycle costs.

All of the appropriate stakeholders, including acquisition, technology, and logistics, are at the table when DAB decisions are made.[7] During the execution of the acquisition process, the Air Force must more fully consider, and plan for, long-term sustainment as a critical component of weapon system acquisition

---

[4]Grover L. Dunn, Director of Transformation, Deputy Chief of Staff for Logistics, Installations and Mission Support, Headquarters United States Air Force. 2011. Expeditionary Logistics for the 21st Century. November 3.

[5]Grover L. Dunn, Director of Transformation, Deputy Chief of Staff for Logistics, Installations and Mission Support, Headquarters United States Air Force. "Expeditionary Logistics for the 21st Century (eLog21)." Presentation to the committee, January 17, 2011.

[6]Rick Conlin and Jim McIntosh. 2010. Collaborative Management Will Improve Weapon System Sustainment. Army Sustainment 42(5):55-59. Available at http://www.almc.army.mil/alog/issues/SepOct10/collaborative_mgmt.html. Accessed March 23, 2011.

[7]DoD. Defense Acquisition Guidebook. Section 10.2, p. 479. Available at https://dag.dau.mil/Pages/Default.aspx. Accessed May 16, 2011.

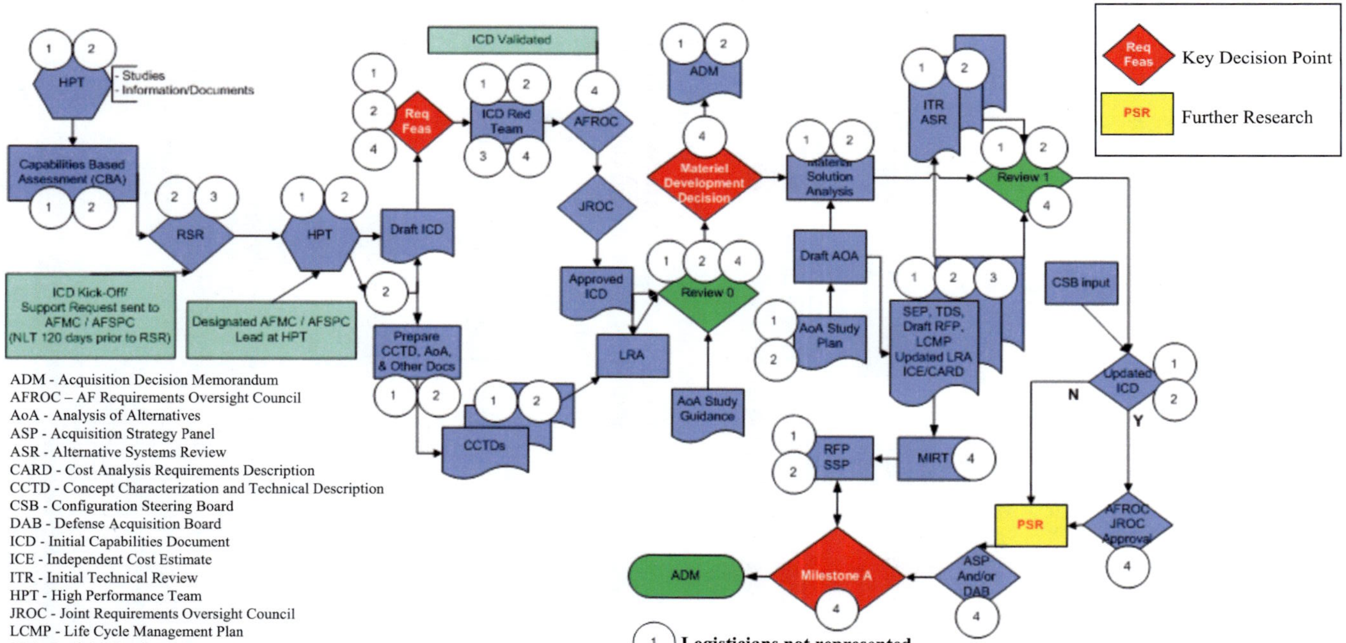

ADM - Acquisition Decision Memorandum
AFROC – AF Requirements Oversight Council
AoA - Analysis of Alternatives
ASP - Acquisition Strategy Panel
ASR - Alternative Systems Review
CARD - Cost Analysis Requirements Description
CCTD - Concept Characterization and Technical Description
CSB - Configuration Steering Board
DAB - Defense Acquisition Board
ICD - Initial Capabilities Document
ICE - Independent Cost Estimate
ITR - Initial Technical Review
HPT - High Performance Team
JROC - Joint Requirements Oversight Council
LCMP - Life Cycle Management Plan
LRA - Logistics Readiness Assessment
MIRT - Multi-functional Internal Review Team
PSR - Program Support Review
RFP - Request For Proposal
RSR - Requirements Strategy Review
SEP - Systems Engineering Plan
SSP - Source Selection Plan
TDS - Technology Development Strategy

1 **Logisticians not represented**

2 **Logistics requirements (enterprise/program) not adequately considered**

3 **Policy Enforcement and Logistics Access**

4 **Enterprise/program Logistics not properly assessed**

FIGURE 3-1
Pre-Milestone A Defense Acquisition Board activities showing inadequate logistics involvement. SOURCE: Grover L. Dunn, Director of Transformation, Deputy Chief of Staff for Logistics, Installations and Mission Support, Headquarters U.S. Air Force. "Expeditionary Logistics for the 21st Century (eLog21)." Presentation to the committee, January 17, 2011.

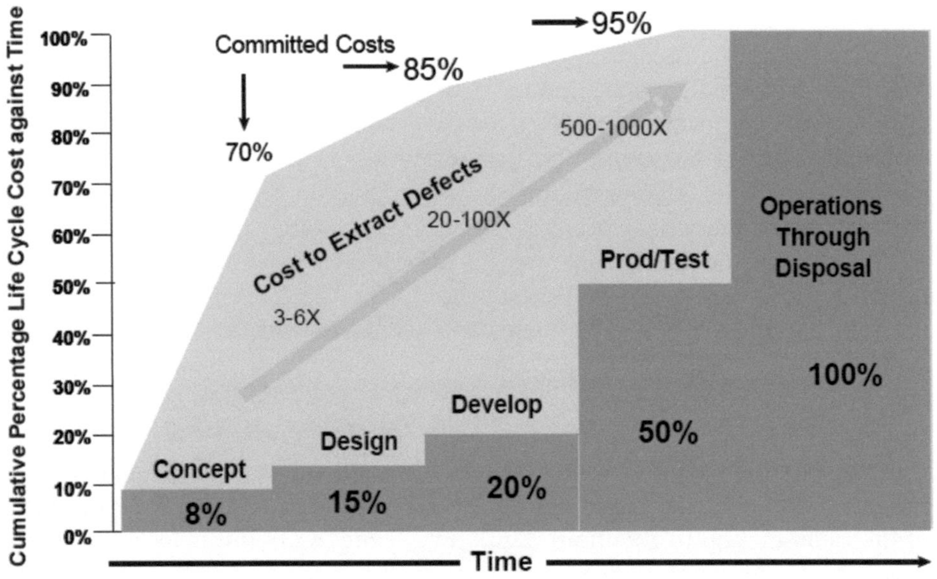

FIGURE 3-2
Life-cycle cost commitment. SOURCE: INCOSE Systems Engineering Handbook Version 3, June 2006.

decisions. Efforts such as LRT will be helpful to moving meaningful sustainment considerations to earlier in the acquisition process; however, the Air Force must strengthen its own leadership to properly maintain a balance between capability and lifecycle sustainment costs.

As described in detail in Chapter 2, the Air Force's sustainment strategy is founded on the concept of developing an Integrated Life Cycle Management (ILCM) enterprise, and it has already established an ILCM Executive Forum. This forum has been charged to ensure that sustainment considerations are fully represented as part of the development process for new weapon systems, as mandated in DoD Instruction 5000.02 and *DoD Weapon System Acquisition Reform Product Support Assessment.*[8,9]

The Air Force's research and development (R&D) activities are conducted in the Air Force Research Laboratory (AFRL), which primarily focuses on technology

[8]To review the DoD Instruction 5000.02, see https://acc.dau.mil/CommunityBrowser.aspx?id=332529. Accessed May 4, 2011.

[9]DoD. 2009c. DoD Weapon System Acquisition Reform Product Support Assessment. November. Washington, D.C.: Office of the Under Secretary of Defense for Acquisition, Technology and Logistics. Available at https://dap.dau.mil/career/log/blogs/archive/2010/01/28/implementation-of-dod-weapon-system-acquisition-reform-product-support-assessment-psa-recommendations.aspx. Accessed November 22, 2010.

for new weapon systems in the concept stage or early development stage. During the course of the study, the committee identified some history of technology transition from the Air Force Research Laboratory to the operational fleets to provide solutions to near-term problems arising from (1) new requirements and (2) upgrades to sustain the fleet readiness and to reduce sustainment costs through improvements in repair techniques. These functions, however, have not been a priority for the Air Force. Chapter 5 describes some of the major technology transfer successes and will discuss how the process and intensity of technology transfer and sustainment-related investments have been insufficient.

Military construction investments and the portion of the O&M budget that supports facility maintenance are vital to the sustainment mission, because adequate facilities are required to support the new and evolving system requirements and to address normal obsolescence and deterioration from use. The Base Realignment and Closure (BRAC) Act that consolidated the Air Logistics Centers (ALCs) provided some modernization and replacement, but the significant ages of some of the facilities require attention. Legislation defines a minimum level of funding to be directed to depot maintenance functions.[10]

> **Finding 3-1.** The Air Force acquisition process has emphasized (1) initial acquisition cost estimates; (2) promised achievement of key performance parameters; and (3) optimistic support costs estimates, without adequately addressing longer-term sustainment considerations that drive most of the total lifecycle costs.

CURRENT RESOURCES AND INVESTMENTS

Through 2007, the Air Force experienced a significant reduction in funding for O&M, a key driver of the sustainment enterprise. Several factors contributed to this reduction, including contingency operations, increasing fuel costs, aging aircraft spares requirements, costs associated with BRAC actions, and lost savings due to congressional restrictions on retirement and divestment of legacy aircraft. Several important factors put pressure on the Air Force's ability to manage an aging fleet. Specifically, from 1988 through 2008, personnel costs increased 57 percent, while personnel end strength decreased 8 percent.[11] Furthermore, the cost to operate aircraft rose 179 percent while aircraft inventories declined. These rising costs put pressure on the Air Force's ability to modernize the fleet because procurement

[10]For additional information on the BRAC and also how the legislation consolidated the ALCs, see http://www.globalsecurity.org/military/facility/brac.htm. Accessed May 2, 2011.

[11]United States Airforce (USAF). 2007. U.S. Air Force FY08 President's Budget. February. Available at http://www.saffm.hq.af.mil/shared/media/document/AFD-070212-012.pdf. Accessed April 15, 2011.

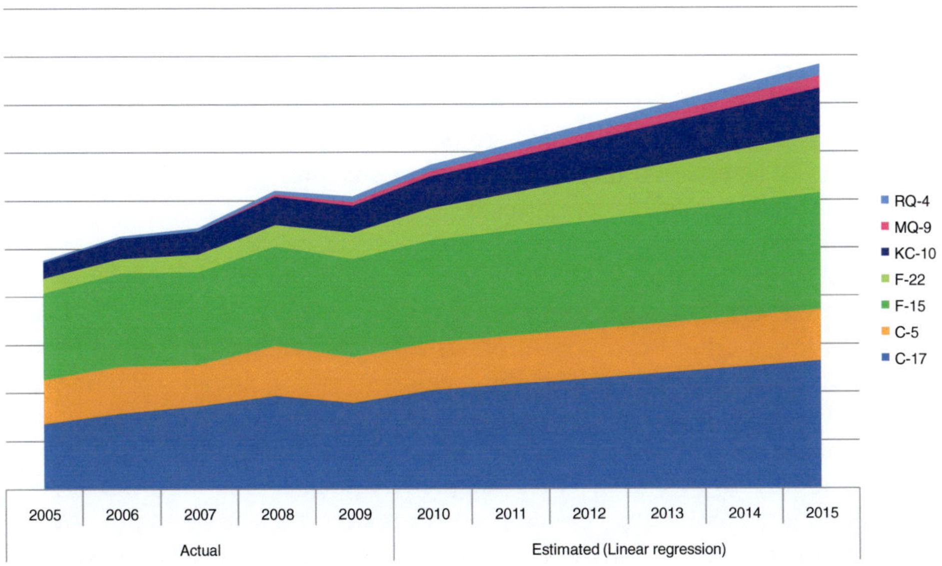

FIGURE 3-3
Weapon system sustainment growth for selected programs. SOURCE: Debra K. Tune, Principal Deputy Assistant Secretary of the Air Force for Installations, Environment and Logistics, Office of the Assistant Secretary of the Air Force for Installations, Environment and Logistics. "Developing the Right Product Support Concepts for the Future." Presentation to the committee, October 20, 2010.

funding as a percentage of Air Force Funding declined 19 percent over a 22-year period.[12] Figure 3-3 highlights this sustainment growth for selected weapon systems. Growth in sustainment costs for these systems is the result of increasing costs and increasing aircraft inventories in some cases.

These funding pressures have left the Air Force with an aging fleet that requires more funds to operate at viable levels that meet aircraft availability targets. However, with the submission of the Fiscal Year (FY) 2012 President's budget, it seems unlikely that the Air Force will receive the resources necessary to significantly improve the viability of its fleet. Further funding reductions will be taken from the sustainment O&M accounts in the form of efficiencies as the Administration works to save $1.1 trillion over the next 10 years to ease deficit concerns. The Secretary of Defense (SECDEF) has outlined the savings proposed by the Military Departments, and the Air Force's proposal totals $34 billion from 2012 to 2016. Some of these savings will come from the sustainment enterprise as indicated by the SECDEF's

[12]Ibid.

statement that the Air Force would "improve depot and supply chain business processes to sustain weapon systems."[13]

> **Finding 3-2.** It is uncertain that the efficiencies envisioned in the FY2012 budget can be achieved. If these efficiencies do not come to pass, sizable impacts to fleet readiness should be expected.

The Air Force Budget for FY2012 totals \$119 billion (the "Blue" budget),[14] which represents a decrease of almost \$600 million from FY2011 levels. The O&M appropriation, a critical component of sustainment funding, decreases by more than \$500 million, despite the increasing requirements for aircraft sustainment. It was widely reported that the Air Force needed an additional \$7 billion to fund 82 percent of the sustainment requirement over the next 5 years. This funding level was not affordable, so the Air Force conducted a total review and prioritization of remaining requirements. The result of this review was a \$4 billion increase in funding, coupled with \$3 billion in efficiencies within the sustainment enterprise, enabling the Air Force to fund 84 percent of the requirement over the next 5 years.

The Air Force weapon system sustainment (WSS) budget for FY2012 is \$9.7 billion, or 69.7 percent of the full requirement. It consists of four primary components: (1) depot maintenance, (2) contractor logistics support (CLS), (3) sustaining engineering, and (4) technical orders. Funding levels (excluding funding for Overseas Contingency Operations), and the percentage of the requirement that is funded, for each of these areas is as follows:

1. Depot Maintenance (\$3.8 billion, 75 percent)—Includes major overhaul and/or rebuild of parts, assemblies, subassemblies and end items, manufacture of parts, technical assistance, software maintenance, and storage.
2. CLS (\$5.4 billion, 69 percent)—Includes contract support for a program, system, training system, equipment, or item used to provide all or part of the sustainment elements in direct support of an approved sustainment strategy to include operations.
3. Sustaining Engineering (\$0.4 billion, 48 percent)—Includes engineering efforts required to review, assess, define, and resolve technical or supportability deficiencies revealed in fielded systems, products, and materials.
4. Technical Orders (\$0.1 billion, 60 percent)—Includes user friendly, techni-

[13]Briefing: "Secretary's Efficiency Initiatives: Follow-up to Jan 6 Speech by Secretary Gates."
[14]United States Air Force FY2012 Budget Overview, SAF/FMB, February 2011.

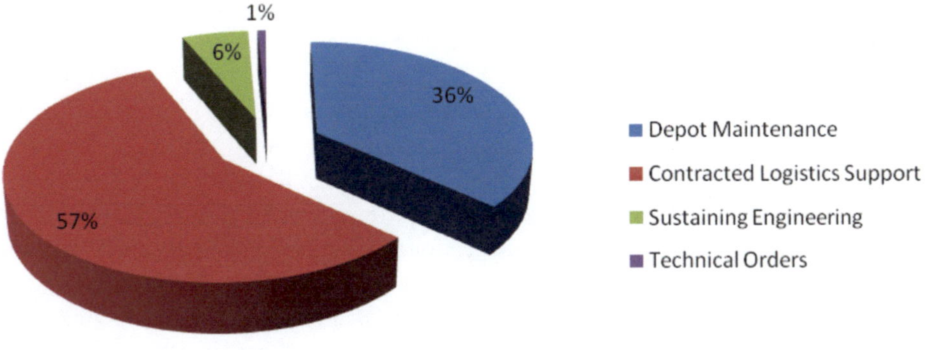

FIGURE 3-4
Components of the Air Force weapon system sustainment portfolio. SOURCE: Data provided by Major General Alfred Flowers, Deputy Assistant Secretary for Budget, Office of the Assistant Secretary of the Air Force for Financial Management and Comptroller.

cally accurate, and up-to-date technical data at the point of use that is acquired, sustained, distributed, and available for all users.[15]

Figure 3-4 highlights the relative amounts of each of the four areas listed above in the WSS program.

Although the Air Force considers these categories to be the components of weapon sustainment, there are other costs that should be mentioned within the context of sustainment. Specifically, the categories of depot-level reparables and consumable supplies should be considered when viewing sustainment in the larger construct. These costs, which are key components of the Air Force flying hour program, comprise an additional $2.6 billion in sustainment costs. Additionally, the cost of labor at the three ALCs totals nearly $2 billion.[16] Adding these costs to the WSS portfolio brings overall sustainment spending in FY2012 to $14.3 billion, nearly 32 percent of the Air Force O&M budget.[17]

---

[15]Scott A. Haines, Colonel, USAF. *Capabilities-based Resourcing for Air Force Weapon System Sustainment.* Air Force Journal of Logistics. Vol XXXIV, Numbers 1 and 2, Annual Edition. Available at http://www.aflma.hq.af.mil/shared/media/document/AFD-101122-029.pdf. Accessed August 18, 2011.

[16]Cost of labor includes pay/salaries. It is not intended to be fully burdened with medical, retirement costs, among other factors.

[17]Department of the Air Force, FY 2012 Budget Estimates, February 2011, Operation and Maintenance, Vol II.

AIR FORCE SUSTAINMENT INFRASTRUCTURE

Infrastructure issues have a direct impact on the ability of the Air Force to sustain its weapon systems. As Figure 3-5 shows in the row entitled "Actual/Estimated Inv," the Air Force investments in depot maintenance production/facilities and equipment were approximately $3 billion over the past 8 years.

The Air Force is actually exceeding the requirements of 10 U.S.C. § 2476, which requires that 6 percent of the annual revenue be invested in the physical facilities and infrastructure that perform the work. Yet the facilities are neither optimized nor in some cases suitable for present and future needed capabilities. For example, 1950s-era engine test stands that are marginally serviceable due to obsolete and nonsupportable instrumentation and fixtures are still being used, and computers in the B-2 Weapon System Support Center Software Integration Laboratory are no longer supported by the original manufacturer or sub-tier vendors. In addition, there is a lack of availability of engine test stands to accommodate the F119 engine

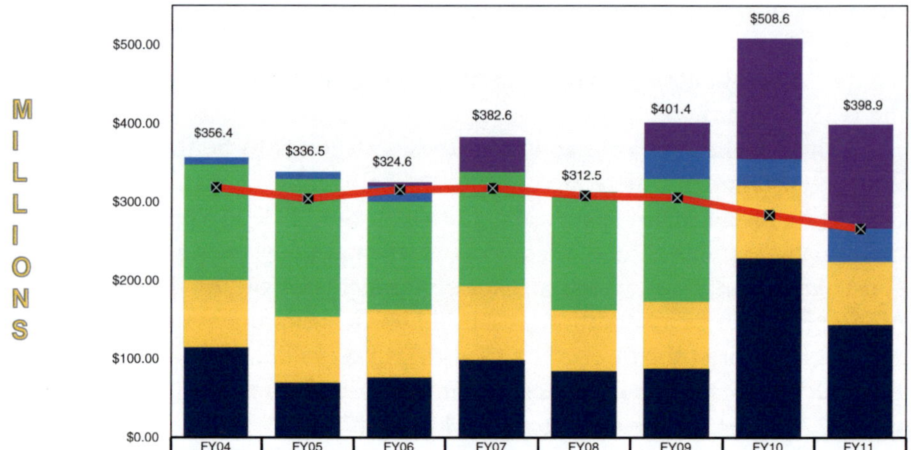

	FY04	FY05	FY06	FY07	FY08	FY09	FY10	FY11
ACFT PROC.	0.0	0.0	4.6	44.6	0.0	36.0	153.3	132.7
MILCON	8.9	7.7	19.8	0.0	0.0	36.0	34.1	41.8
DMT	147.8	175.5	137.3	145.2	150.1	156.2	0.0	0.0
M&R	85.0	83.8	86.3	94.1	77.8	85.2	92.5	80.7
CPP	114.7	69.5	76.6	98.7	84.6	88.0	228.7	143.7
REQ. INVEST.	318.2	304.1	315.7	317.5	308.0	305.7	283.6	265.9
ACTUAL/ESTIMATED INV.	356.4	336.5	324.6	382.6	312.5	401.4	508.6	398.9

FIGURE 3-5
Investments in the Air Force maintenance infrastructure. SOURCE: Debra K. Tune, Principal Deputy Assistant Secretary of the Air Force for Installations, Environment and Logistics, Office of the Assistant Secretary of the Air Force for Installations, Environment and Logistics. "Developing the Right Product Support Concepts for the Future." Presentation to the committee, October 20, 2010.

for the F-22, the F117 engine for the C-17, and the engines for the planned new tanker fleet.

There is tremendous capability at the ALC facilities. In some cases the capabilities exceed what is expected for local maintenance requirements and border on full-scale manufacturing. Some of this is needed, but in a modern facility with modern digitalized technologies, redundancy in the sense of manufacturing capability is not widely needed. Improved planning in response to requirements, flexible manufacturing concepts with a focus on maintenance and repair capabilities, and integration of appropriate repair and maintenance technologies would allow the Air Force to find efficiencies and improve costs.

More than 100 aircraft were in for some type of maintenance or repair at Warner-Robins Air Logistics Center (WR-ALC).[18] Significant numbers were being worked outdoors, where they were subjected to numerous weather impacts. In addition, the aircraft were affected by a less-than-optimal production environment because of distance from parts and lighting, among other factors. These conditions do not generally exist in the top-performing echelons of industry but seem to be taken for granted by the Air Force to meet production needs. The workforce accomplishes the work despite the impediments and possesses a terrific "can do" attitude; however, the work is not necessarily efficient, and resources may not be effectively allocated to meet the work volume.

There is significant degradation of base infrastructure. Aging facilities have potential for catastrophic loss of heating, cooling, power, and other utilities systems that are essential for production. Although these disruptions occur from time to time at all Air Force installations, in the depot production environments, where work center revenue accrual may be in the hundreds of thousands of dollars per day and millions over the center, the impact is real, immediate, and measurable.

Investment in Infrastructure

The Ogden (OO-ALC), Oklahoma City (OC-ALC), and WR-ALCs were established in 1940, 1941, and 1943, respectively.[19] Over the years, the ALCs have expanded their support to address the changing needs of the military. The ALCs provide support to long-standing platforms as well as to newer platforms that contain advanced technologies. Newer methods of repair and maintenance are often required to support these newer platforms. Reorganization of the maintenance and test areas is required to provide efficiency in returning the platforms to service. The facilities must be upgraded to ensure effective service.

[18]The committee visited WR-ALC on January 5-6, 2011.

[19]For more information, see http://www.hill.af.mil/library/factsheets/factsheet.asp?id=5830. Accessed May 16, 2011.

TABLE 3-1 Comparison of the Needed Maintenance Costs Versus the Budgets

	Millions of Dollars	
Base	Needs	Budget
Tinker[a]	>985	32-55
Odgen[b]	>800	4
Warner-Robins[c]		4-5

[a]Floyd Craft, Director, 547th Propulsion Maintenance Squadron. "Engine Test Cell Strategy." Presentation to the committee on January 12, 2011.

[b]Major General Andrew E. Busch, Commander, OO-ALC. "Ogden ALC Mission Briefing." Presentation to the committee on January 31, 2011.

[c]Brigadier General Lee K. Levy, II, Commander, 402nd Maintenance Wing. "402 MXW Perspective to the Air Force Studies Board." Presentation to the committee on January 6, 2011.

Available funds are not sufficient to keep up with the needs of maintaining, repairing, and updating old facilities and for providing new capabilities. At the 402d Maintenance Wing at WR-ALC, the average building age is 29.4 years with the oldest building being 68 years. At OC-ALC the buildings date back to the 1940s, and many of the original systems are still in place. A comparison of the cost of needed maintenance versus the budgets is shown in Table 3-1.

The maintenance needs shown in Table 3-1 illustrate the disparity that exists between the resources (budget) available for infrastructure maintenance and the perceived needs for infrastructure maintenance. These requirements are those that are foundational to maintaining infrastructure and include plumbing, electrical, and HVAC systems as well as the pavements, buildings, and test stands. There are concerns about catastrophic events in which failure of any of these systems could have an adverse impact on the bases' missions. Insufficient maintenance and upgrades can result in failures that are not easily repaired, resulting in delays to making airplanes available. Although OC-ALC has leased an updated, former General Motors facility, it is used primarily for manufacturing, component repairs, and parts replacement and not for testing. The test facilities are old and cannot be utilized with the newer platforms that are a part of the future plans of the base. The facilities' layouts are also an important consideration. Although the ALCs have applied some amount of lean practices to optimize rate and flow, the facilities must be correctly aligned to allow for this optimization.

The bases have accepted more and different work because of movement of work from other bases, but this desire for flexibility creates challenges, because current facilities that are inadequately maintained can be taxed by new test fixtures, equipment, and methods. Newer platforms often contain new technologies, and the facilities must be updated to accommodate repair and maintenance technologies that can support these platforms. This may not be possible with the current facilities because of the aged power, plumbing, and other systems. In short, the investment in facilities maintenance is not adequate to accommodate current workloads and to address future needs.

Ground Equipment

Ground support equipment typically includes all implements, tools, and devices (mobile or fixed) required to inspect, test, adjust, calibrate, appraise, gage, measure, repair, overhaul, assemble, disassemble, transport, safeguard, record, store, or otherwise function in support of a platform, either in the research and development phase or in an operational phase. Different equipment is used to assess, repair, or provide maintenance, and test for the quality of operational outcomes. Although the ground equipment tends to be old, it is adequate to provide these functions for current needs. However, the equipment will not be adequate to perform new work that is planned for the ALCs, such as an increase in new testing for the C-17 at OC-ALC.[20]

> **Finding 3-3.** Several critical plant and equipment investments will be needed in the near future. Without these investments, the Air Force will not be able to fully support current and future organic workloads, and thus will face longer periods of CLS with the inherent 10 U.S.C. § 2466 ramifications.

> **Recommendation 3-1.** The Air Force should continue funding depot plant and capital equipment and, at the same time, be guided by focused analyses to ensure that constrained funding is provided to the most critical sustainment needs to avoid future support impacts and to meet 10 U.S.C. considerations.

CURRENT AIR FORCE SUSTAINMENT PROCESSES

The following section provides a broad review of Air Force sustainment processes. Some of these processes are addressed in detail elsewhere in this report. Covered in this chapter, however, are workforce,[21] acquisition, the supply chain, maintenance processes, resourcing efforts, Fleet Viability Board efforts, logistics

[20] Floyd Craft, 547th Propulsion Maintenance Squadron director. "Engine Test Cell Strategy." Presentation to the committee on January 12, 2011.

[21] Workforce is defined in this report as the policies, procedures, and issues affecting the development and retention of the knowledge and skill sets needed by the labor force to maintain existing systems and to be ready to maintain future systems.

support processes, Expeditionary Logistics for the 21st Century (eLog21), and obsolescence and diminishing manufacturing sources.

Workforce

This section discusses knowledge and skill sets, engineering staff focus, and personnel allocation. The ALCs expressed concern about present-day workforce knowledge and skill sets and the ongoing retirements of many senior employees. At one of the ALCs, the average employee has only 5 years on the job.[22] Across the sustainment enterprise and particularly at the ALCs, a number of active recruitment and knowledge retention actions are ongoing, such as recruiting personnel from nearby career and technical education (CTE) schools, arranging with state technical training agencies, and chartering senior employees to mentor junior employees. In this way, long-term skills and experience are passed on to supplement and fuse with new techniques/approaches learned at the CTE schools.

Harvesting and maintaining knowledge and lessons learned might be further improved by making information-sharing systems, such as SharePoint, available on the shop floor. Knowledge accumulation, editing, and distribution would, of course, need to be addressed when implementing these systems. Continued use of formal training programs is another useful investment, especially as technology insertions occur. These observations are consistent with the Air Force Maintenance Strategic Plan.[23]

At the same time, there are real concerns with the evolution of CLS platforms to organic support, such as where the technical workforce will come from an era of constrained workforce levels and new technology introduction. The belief is that the workforce applied to current legacy systems will easily transition to the newer platforms or support concepts. This may be more theory than reality in practice. In addition to airframe and other logistics issues, questions exist regarding software sustainment over the lifetime of a weapon system.[24]

> **Finding 3-4.** At the current time, the ALCs are doing an adequate job of personnel recruitment and knowledge retention. However, long-term concerns exist as retirements increase and systems move from CLS to organic support.

Equally important to the overall sustainment activities is the engineering staff for depot maintenance support. A common issue for the three ALCs relates to

---

[22]Major General Bruce A. Litchfield, personal communications to committee members on January 11, 2011.

[23]USAF. 2008 Air Force Maintenance Strategic Plan.

[24]Software sustainment is discussed in considerable detail in Chapter 4.

engineering support for the depot maintenance repair lines. Engineering support is provided by the program offices and the commodity management offices. The engineering staff in these offices has conflicting priorities between depot support, evaluations of field requests, requests for manufacturing first article tests, and so on. In fact, engineering disposition for issues on the aircraft, engine, and commodity repair lines is delayed from time to time with the attendant impact to production schedules. Engineering resources are constrained as part of the O&M-funded manpower baseline and often don't reflect workload needs. Interestingly, the United States Navy reported that these issues are nonexistent at its Fleet Readiness Centers.[25] Although the Navy's engineering staff supports the same type of function, the leadership recognizes and prioritizes support to the maintenance operations.

As manpower resources were constrained over the past several years, the Air Force eliminated Combat Logistics Support Squadrons at the ALCs, an action with unintended consequences. These squadrons provided deployable military maintenance specialists to perform depot-level tasks at field locations. Since the squadrons' elimination, qualified civilian rather than Air Force military craftsmen are now deployed to perform the depot-level tasks. One depot maintenance supervisor explained that this change has created her most pressing challenge: although she has a full complement of specialists, many are temporary or contract employees who have backfilled the skilled workforce. The learning curve for these new employees has been high, and times to complete tasks have been higher than they should be, which has routinely impacted the schedule.[26]

Finding 3-5. The ALCs have a great demand for engineering support. At the same time, the engineering staffs have conflicting priorities.

Recommendation 3-2. The Air Force should establish clear priorities for engineering activities and consider examining lessons learned and the applicability of the Navy model of workforce issues.

Acquisition

A key message repeated by Air Force officials is that sustainment must be considered at the outset of the acquisition life cycle. During the acquisition process, a systems view of the platform with a definition of the total life cycle is required to accurately reflect realistic requirements. Lessons learned from current plat-

[25]Captain Fred Melnick et al., Fleet Readiness Center Southwest, Roundtable Discussion, March 29 2011.

[26]OC-ALC program managers, personal communications with committee members on January 11, 2011.

forms can increase accuracy of predicted sustainment needs relative to scheduled maintenance. Moreover, documented findings and trends related to aging aircraft, condition-based maintenance, exposures to extreme environments, and product disposal can provide a more accurate picture of the processes and costs related to the acquisition process. As pointed out earlier in this chapter, Figure 3-1 highlights the gaps with respect to the consideration of logistics in the early planning of programs. Further, long-term sustainment needs are often minimized early in the acquisition process in favor of performance or shorter-term financial considerations. Early involvement of sustainment personnel at multiple points before Milestone A will allow for more accurate planning and budgeting of sustainment needs.

Finding 3-6. Currently, sustainment resource planning is not adequately planned and budgeted for during the acquisition process.

Supply Chain

From the purest standpoint, the supply chain refers to the complete cycle of acquiring the raw material to produced goods; the manufacturing, stocking, and storage of the produced goods; transportation to locations of needs; resupply efforts; maintenance activities on these items; disposition after consumption; condemnation; and so on. In this particular discussion, the supply chain is confined to the spare parts available to users and to the approaches for obtaining products and services (i.e., contractor-operated or organic processes).

The organic supply chain processes are those activities that are supported by the Air Force and the DoD, including the ALCs and the Air Force Global Logistics Support Center (AFGLSC). The AFGLSC is the principal manager of the Air Force supply chain, and the Defense Logistics Agency is organic to DoD. Many organizational and process changes have been associated with these organizations over the past six years. These changes and the degrees of success are covered in considerable detail in Chapter 4 because they directly involve resourcing of the ALCs.

The committee observed several examples of very effective and reasonably efficient contractor-operated supply chains. For example, in its partnership with Boeing for C-17 lifecycle support, the Air Force has retained limited core depot maintenance capability and maximized the use of the contractor support and supply system, which has led to a significant reduction in cost per flight hour as highlighted by Figure 3-6.

Such partnerships are in line with DoD Instruction 5000.02: "Support Concepts for new and modified systems shall maximize the use of contractor provided, long term, total life logistics support that combines Depot Level maintenance along

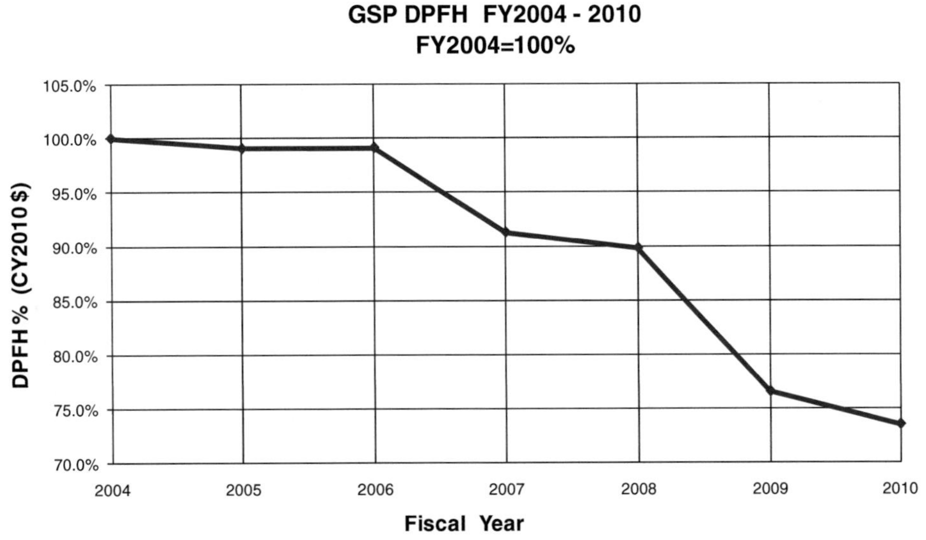

FIGURE 3-6
Dollars per flight hour (DPFH) for the C-17 FY2004-FY2011. SOURCE: Gustavo Urzua, The Boeing Company. Personal communication with the committee on May 19, 2011. NOTE: The data portrayed are based on the Boeing contract with the Air Force and include profit and full award fee and are a performance-based logistics yearly contract for labor and material.

with wholesale and selected materiel management functions."[27] The Air Force and the contractor have a shared infrastructure that reduces organic sustainment footprints, promotes Air Force partner access to the international C-17 infrastructure, generates opportunities for robust exchange agreements (spares, equipment, Tech Data), and minimizes the size and cost of spares pools.

Finding 3-7. Collaborative partnerships with contractor-operated supply chains have resulted in improved efficiencies and lowered costs.[28]

Maintenance Processes

Maintenance processes consist of field and depot activities that allow the platforms to be ready for service. Maintenance activities include scheduled maintenance as well as unplanned repair or replacement activities. Both scheduled and

[27]DoD. 2008. *Operation of the Defense Acquisition System.* December 8. http://elastic.org/~fche/mirrors/www.jya.com/dodi/dodi-5000-02.pdf. Accessed August 18, 2011.

[28]This topic is discussed extensively in Chapter 4.

unscheduled maintenance can be diverse and complex because of the age and variety of platforms and the conditions to which these platforms are exposed. Maintenance activities can range from simple to complex replacements/repairs to complete fabrication, manufacturing, and installation of complex parts. Maintenance is a key element to the aircraft availability metric.

Challenges to maintenance include inadequate planning, part shortages, and increased turn time, all of which can occur with planned maintenance and thus necessitate unscheduled work and inefficiencies. However, these challenges can be exacerbated by conditions such as aging platforms, unplanned damage, and new technologies that can result in discoveries of new and/or unexpected types of damage requiring extensive or yet-to-be identified technologies to be repaired or maintained.

Additionally, a lack of synchronization between field and depot activities, programmed depot maintenance cycles, stove-piped processes, and general maintenance execution can contribute to not attaining the target aircraft availability. A common and centralized data collection system would allow the ALCs to standardize processes and share technical data, best practices, and lessons learned, which would help to streamline and standardize processes for maintenance and repair.

> **Finding 3-8.** There is no centralized database to document technical activities, best practices, and lessons learned to enhance knowledge management and process improvement.

High Velocity Maintenance (HVM) is one example of Air Force efforts to improve depot maintenance processes. The HVM methodology has been considered and adopted in some areas to improve maintenance performance, reduce flow days, and reduce field-level maintenance. Additionally, the benchmarking of commercial companies and the application of best practices and lean processes are being implemented to streamline maintenance activities. Different lean tools are applied as needed to decrease the time to delivery.[29,30]

Resourcing Processes

The Air Force has a complex, thorough, and highly participative resourcing process for funding its requirements. Following the direction and guidance of the Office of the Secretary of Defense (OSD), the Air Force assembles its current

---

[29]Steve Walker. "High Velocity Maintenance (HVM)." Presentation to the committee, January 12, 2011.

[30]Captain Tyler B.L. Schroder. "F-22 High Velocity Maintenance Program Review." Presentation to the committee, January 7, 2011.

and future budgets in a Future Years Defense Program (FYDP) that summarizes resources (funding, manpower, and forces) as approved by the Secretary of the Air Force (SECAF) and OSD. The FYDP also reports Planning, Programming, Budgeting, and Execution decisions and funding for the next six fiscal years.

The Air Force programming process is essentially the first step to assembling the information and resources necessary to create a budget. The resources and force structure are aligned in a Program Objective Memorandum (POM) that reflects Air Force priorities. At this point in the process, all fiscal resources are fungible, which means they can be realigned between programs and appropriations as needed to fund priorities as outlined by Air Force leadership. During the course of this study, the committee learned of issues and challenges related to the "color of money," that is, not having the money in the right appropriation for spending on a project or program. Although this misappropriation may be a constraint during the fiscal year of execution, it can be largely mitigated during the POM process. Funding issues that arise during the budget year of execution can often be resolved by the Air Force reprogramming process, whereby the Service seeks congressional approval to realign resources between appropriations to fund critical high-priority shortfalls, thereby resolving many color of money issues.

Within the WSS enterprise, the Air Force has implemented a Centralized Asset Management (CAM) system that consolidates resources previously managed by the operational commands. Fiscal resources have been realigned from the operational commands to Air Force Materiel Command (AFMC), are managed at the enterprise level, and include the funding for spare parts, depot-level maintenance, sustaining engineering, technical orders, and aviation fuel.

The CAM program offers several benefits to AFMC and the Air Force. Prior to its implementation, major commands were often faced with difficult decisions related to O&M funding shortfalls in base support, communications, real property maintenance, and many other areas of base activity. As a result, commands would often defer inducting aircraft or engines, previously funded for depot work, and elect to spend those resources on other high-priority shortfalls. Centralizing this funding now provides AFMC with the ability to more effectively manage WSS, scheduling, and work flow. Over time, this process will improve the sustainment levels within the Air Force.

Fleet Viability Board

The Fleet Viability Board provides technical assessments of aging aircraft to both the SECAF and the Chief of Staff of the Air Force (CSAF). Viability is defined as "It can do what we need it to do, when we need it to do it, at a price we are willing to pay."[31] The metric used is the cost per available aircraft or the sum of the modification costs plus the operation and support (O&S) costs per available aircraft. The Fleet Viability Board does not believe that current sustainment investments, infrastructure, and processes are adequate to sustain aging legacy systems and their support equipment because of competition for funding and the short-term view of sustainment. It has recommended that changes be made to contract types to ensure visibility of costs and the deployment of health monitoring technologies.

Logistics Support

Of the many processes that are used to sustain Air Force weapon systems, logistics support is overarching and vitally important, but often under-recognized. In part, logistics support includes the processes associated with engineering data, engineering drawings, technical orders, spare part provisioning, cataloging, and others. Without the successful execution of these and other activities, the WSS would grind to a halt. The committee placed logistics support in the category of too important to ignore but too detailed to cover in depth under the terms of this report. However, the committee recognizes that it is entwined in WSS. The decisions that occur very early in a weapon system's life cycle affect logistics support processes, such as re-procurement, day-to-day supply, and maintenance support, over the remainder of the system's life cycle and determine success or failure as well as costs. Logistics support processes are precisely among the key processes that are impacted when decisions are made for various support concepts. Consequently, when decisions makers arrive at major milestones, the sub-elements of logistics support must be considered. Simply put, early decisions should be made with full understanding of their long-term costs and implications.

eLog21

eLog 21 is a transformational campaign aimed to drive improvements to Air Force logistics support and WSS. This umbrella effort consists of multiple logistics transformation initiatives and primarily aims to increase equipment availability and reduce operations and support costs. It addresses the fact that future budgets will remain flat or decrease, resulting in calls for more "efficiencies." It relies heavily on process tools such as LEAN and Six Sigma and introduces a Logistics Enterprise Architecture (LogEA) as a roadmap, which provides an authoritative source to define both operational and systems approaches to Air Force logistics. eLog21 defines and aligns the organizational vision, mission, goals, objectives, and *processes*

[31] Fran Crowley, Director, Air Force Fleet Viability Board. "AF FVB Feedback for AF Studies Board." Presentation to the committee on December 7, 2010.

with information technology initiatives. It established the aforementioned ILCM Executive Forum and aids in the integration of Air Force acquisition and logistics policy. It aims to ensure that life-cycle logistics are addressed at every step from the lab, to the requirements, to the design and testing, to the manufacturing and delivery process. It explores innovative technologies and incorporates a number of product support initiatives.

In addition to the LogEA roadmap, the Expeditionary Combat Support System (ECSS) is a key underpinning of eLog21. ECSS represents a standardization of sustainment processes at every level and uses an Enterprise Resource Planning tool to implement and enforce the approved rule set. ECSS is discussed further in Chapter 4, but it is worthwhile to note here that the Air Force estimates that 240 legacy systems will be replaced by ECSS. The Air Force has spent $897 million through FY2010 on ECSS, expects to spend an additional $2.71 billion on investment through FY2017, and will see a total life-cycle cost of $5.1 billion through FY2027.[32] ECSS is clearly a major Air Force investment in the future.[33] Phase I of ECSS development and deployment has been challenging, and the program is at risk, but the need for ECSS or its equivalent is undeniable.

> **Finding 3-9.** eLog21 shows promise in driving improvements to Air Force logistics support and WSS. The campaign deserves full visibility and should be the direct responsibility of the sustainment commander.[34]

Obsolescence and Diminishing Manufacturing Sources

Obsolescence and diminishing manufacturing sources are two conditions that can affect the production and sustainment of the platforms. Obsolescence is defined as the process or condition by which a piece of equipment becomes no longer useful. It can also mean that the form and function are no longer current or available for production or replacement/repair. Introduction of new technologies may cause older technology to become less supportable because of diminished availability of parts and suppliers. Obsolescence must be resolved before the process of dealing with Diminishing Manufacturing Sources and Material Shortages (DMS/MS) occurs, especially if the system is still in production.

DMS/MS is loss or impending loss of the last known manufacturer or supplier

---

[32]Grover L. Dunn, Director of Transformation, Deputy Chief of Staff for Logistics, Installations and Mission Support, Headquarters U.S. Air Force. Personal communication to the committee, May 23, 2011.

[33]Government Accountability Office. 2010. DoD Business Transformation: Improved Management Oversight of Business System Modernization Efforts Needed. October. Available at http://www.gao.gov/products/GAO-11-53. Accessed May 16, 2011.

[34]eLog21 is discussed in greater detail in Chapter 2.

of raw materials and other critical components for production or repair/replacement parts. As the weapon systems' service lives are extended, the loss of supply chain and manufacturing capability are an increasing concern. During the course of the study, Air Force leaders repeatedly emphasized the importance of managing obsolescence and DMS/MS.[35,36]

Ultimately, both obsolescence and DMS/MS will result in the inability to provide parts or components to the weapon systems, which poses a risk to readiness. Performance Based Logistics (PBL) is an important obsolescence mitigation strategy. DoD Instruction 5000.02 states, "The PM shall employ effective Performance-Based Life-Cycle Product Support (PBL) planning, development, implementation, and management. Performance-Based Life-Cycle Product Support represents the latest evolution of Performance-Based Logistics. Both can be referred to as 'PBL.' PBL offers the best strategic approach for delivering required life cycle readiness, reliability, and ownership costs."[37] Because this process is performance-based and focuses on weapon system availability and lowering costs, it can be accomplished organically, through suppliers, or a combination thereof. PBL tackles the problem of aging by instituting incentives between the government and the weapon system manufacturer to ensure that support providers continuously modernize and improve their systems and methods of support.

CONCLUDING THOUGHTS

This chapter has addressed the sustainment investments, infrastructure, and processes that are currently in place. Investments in sustainment have not met the challenges of supporting an aging, highly stressed fleet. Current acquisition practices do not consider sustainment early enough in the planning process. The ALCs function adequately, but they are not optimized for current and future needs. Investments in infrastructure, personnel, and ground equipment are needed for optimization. Activities such as eLog21 and HVM are aimed at improving processes for maintenance.

[35]Sue Lumpkins, Deputy Director of Logistics, Office of the Deputy Chief of Staff for Logistics, Installations and Mission Support, Headquarters United States Air Force. "Air Force Studies Board Sustainment Study." Presentation to the committee, October 20, 2010.

[36]Blaise J. Durante, Deputy Assistant Secretary for Acquisition Integration, Office of the Assistant Secretary of the Air Force for Acquisition. "Budgetary Considerations Related to Sustainment." Presentation to the committee, October 21, 2010.

[37]DoD. 2008. Operation of the Defense Acquisition System. December 8. http://elastic.org/~fche/mirrors/www.jya.com/dodi/dodi-5000-02.pdf. Accessed August 18, 2011.

4

Assessment of Air Force Air Logistics Centers

INTRODUCTION

This chapter addresses element 4 of the terms of reference (TOR): "Determine if the Air Logistics Centers have the necessary resources (funding, manpower, skill sets, and technologies) and are equipped and organized to sustain legacy systems and equipment and the Air Force of tomorrow." The U.S. Air Force (USAF) currently has three Air Logistics Centers (ALCs), operating under the Air Force Materiel Command (AFMC), which provide acquisition, modification, and maintenance support for the Air Force aircraft fleets, end items, commodity parts, and some missile systems.

The ALCs are complex, multi-faceted organizations. They provide support to the Air Force and other components of the Department of Defense (DoD) on numerous product lines. As shown in Figure 4-1, the Warner-Robins Air Logistics Center (WR-ALC), founded in 1943 and located on Robins Air Force Base, Georgia, serves as the primary modernization, sustainment, and depot maintenance center for a variety of aircraft, including the U-2, C-5, C-17, all models of the C-130, E-8, and F-15, and other important aircraft.[1,2] WR-ALC also has the Air Force's primary

[1]Major General Robert H. McMahon, Commander, Warner-Robins Air Logistics Center (WR-ALC). "Commander's Briefing." Presentation to the committee, January 6, 2011.

[2]United States Air Force (USAF). 2001. "A Brief History of WR-ALC and Robins AFB." September 1. Available at http://www.robins.af.mil/shared/media/document/AFD-070226-039.pdf. Accessed March 22, 2011.

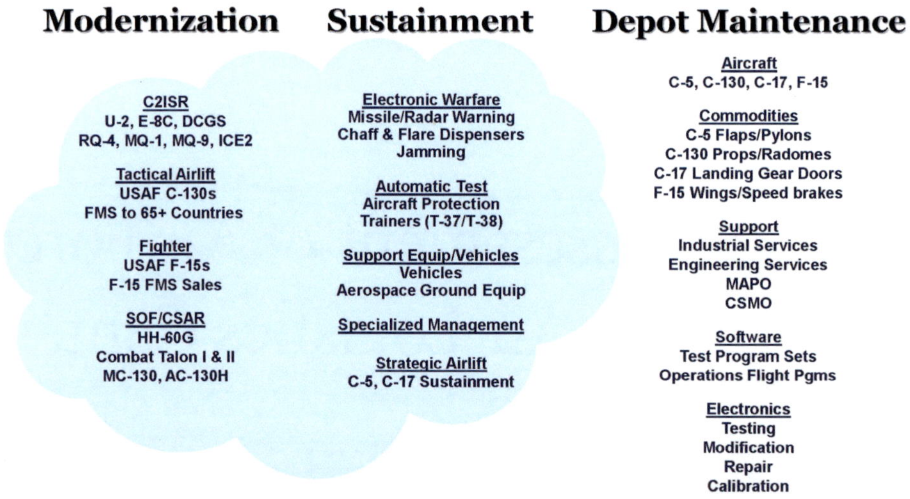

FIGURE 4-1
Warner-Robins Air Logistics Center (WR-ALC) operation areas of responsibility. SOURCE: Major General Robert H. McMahon, Commander, WR-ALC. "Commander's Briefing." Presentation to the committee, January 6, 2011.

responsibility for avionics systems management, support equipment management, and electronic warfare systems and the significant maintenance of these systems.

The Oklahoma City Air Logistics Center (OC-ALC), founded in 1941 and located on Tinker Air Force Base, Oklahoma, manages an inventory of more than 2,000 aircraft including the B-1, B-2, B-52, C/KC-135, E-3, VC-25, VC-137, Cruise missile inventories, and 25 other Contractor Logistics Support aircraft.[3,4,5] Additionally, OC-ALC is responsible for all Air Force propulsion systems and management and a sizable portion of the propulsion systems maintenance. Figure 4-2 depicts the core expertise areas of OC-ALC's 76th Maintenance Wing.

As shown in Figures 4-3 and 4-4, the third and last ALC, Ogden Air Logistics Center (OO-ALC), founded in 1940 and located on Hill Air Force Base, Utah, provides weapon system management and sustainment for numerous platforms, including the A-10, F/QF-4, F-16, T-38, A-37, F-4, F-5, F-16, T-37, F-22A, QF-16,

[3]Major General P. David Gillette, Jr., Commander, Oklahoma City Air Logistics Center (OC-ALC). "OC-ALC Strategic Goals." Presentation to the committee, January 11, 2011.

[4]USAF. Undated. Oklahoma City Air Logistics Center Factsheet. Available at http://www.tinker.af.mil/library/factsheets/factsheet.asp?fsID=8552. Accessed March 22, 2011.

[5]Tinker Education and Development website, Force Development Division, OC-ALC.

FIGURE 4-2
Oklahoma City Air Logistics Center (OC-ALC), 76th Maintenance Wing (76 MXW). SOURCE: Caysie Mercer, OC-ALC. "Tinker Today." Presentation to the committee, January 11, 2011.

and BQM-167A.[6,7] Like WR-ALC and OC-ALC, OO-ALC has key areas of specialization, such as the management and repair of Air Force landing gear systems and sustainment management for conventional munitions and the Minuteman Intercontinental Ballistic Missile system.

To establish the degree of complexity, it is important to note that responsibility for program management and sustainment for some of these platforms are shared with the Aeronautical Systems Center (ASC) and the Electronic Systems Center (ESC). For example, the F-22 program office is at ASC, and many sustainment actions are accomplished at OO-ALC. The E-3 and the E-8 have program offices at ESC.

The following sections address the resourcing (i.e., funding, workforce, skill sets, and technologies), equipping, and organizing of the three ALCs to sustain legacy weapon systems and equipment. The term "manpower" is used in the TOR,

[6] Major General Andrew E. Busch, Commander, Ogden Air Logistics Center (OO-ALC). "OO-ALC Overview—Core Competencies & Priorities." Presentation to the committee, January 31, 2011.

[7] USAF. 2010. Ogden Air Logistics Center Factsheet. Available at http://www.afhra.af.mil/factsheets/factsheet.asp?id=16630. March 31. Accessed March 22, 2011.

Current Workload

Weapon system management for
- US: A-10, F/QF-4, F-16, T-38
- FMS: A-37, F-4, F-5, F-16, T-37, Others
- Munitions, Range Threat, Shelters, Telecom, PCCIE, AEWS, TARS

Sustainment support for
- F-22A, QF-16, BQM-167A, Space & C3I, GTACS, Mission Planning, Weather

Specialized Management
Logistics planning for F-35

AEROSPACE SUSTAINMENT DIRECTORATE
OGDEN AIR LOGISTICS CENTER

Future of the Directorate

Weapon system management for
- US: A-10 (2030) F/QF-4 (2016)
  F-16 (2024) T-38 (2020)
  F-22 F-35

Sustainment manager for
QF-16, BQM-167A, F-35?, UAS?

Major Challenges

Aging aircraft
- Fatigue cracks
- Corrosion
- Life-limited/overage parts
- Diminishing manufacturing sources

Migration to new sustainment tools (e.g., enterprise information systems)

FIGURE 4-3
Weapon system management at Ogden Air Logistics Center (OO-ALC). SOURCE: Major General Andrew E. Busch, Commander, OO-ALC. "OO-ALC Mission Briefing." Presentation to the committee, January 31, 2011.

Current Workload

Aircraft maintenance/missiles
- A-10, F-16, F-22, C-130, F-4, MMIII

Software; development/sustainment

Exchangeables
- Landing gear, power systems, generators, electronics, avionics, composites, shelters

Reclamation and storage (AMARG)

309TH MAINTENANCE WING

Future Workload

Aircraft Maintenance
- F-35, MQ1, MQ9
- RCS Verification

Exchangeables
- Landing gear, power systems, composites, generators, F -22, and F -35 associated electronics and software

Reclamation and storage (AMARG)

Major Challenges

ICBM and nuclear related material
- Transition nuclear materiel handling/ management from DLA to AF

Aircraft maintenance
- Aging aircraft structures
- Diminishing manufacturing sources

Data Rights
- Tech Data, Source Code

FIGURE 4-4
Depot maintenance at OO-ALC. SOURCE: Major General Andrew E. Busch, Commander, OO-ALC. "OO-ALC Overview—Core Competencies and Priorities." Presentation to the committee, January 31, 2011.

but this term often equates solely to people. Thus, the term "workforce," which includes not only numbers but also skill sets, education, and training, is used in this chapter.

METHODOLOGY FOR THE ASSESSMENT

The committee assessed the ALCs' activities and resourcing using a variety of inputs, including: (1) briefings at Headquarters Air Force; (2) briefings at Headquarters AFMC; and (3) visits to the three ALCs with presentations and tours by the respective ALC management chains. During the visits to the ALCs, the committee divided into smaller subgroups composed of subject matter experts that aligned with the activities at the specific ALCs. The committee formed an opinion of the ALCs' activities and conditions from detailed presentations and from candid discussions with the workforce. To assess the ALCs' resourcing, the committee required an understanding of how the ALC leadership teams view the current situation and how the Headquarters AFMC and Headquarters Air Force view both the past investment utilization and the current flow of funding to support ALC operations. Consequently, a holistic perspective was taken of the past, present, and future outlooks for assessing resourcing investments.

The format of each ALC visit was guided by an agenda established by the ALC commander in response to a general request for information. Each ALC visit consisted of a combination of briefings, facility tours, and in-depth discussions with the participants. Each ALC was represented by its commander, with the attendance of both senior military and civilian leadership. The briefings were both detailed and comprehensive, the tours were open and thorough, and the discussions were frank and responsive to the questions posed. In addition, the reviews of the activities of Headquarters AFMC and the ALCs focused on the adequacy of the ALCs' resources in terms of organizational structure, responsibilities, funding, workforce, skill sets, and technologies and their current and planned equipment to sustain legacy and future systems.

In addition to the three ALCs, two Air Force major command customers of the ALCs, specifically the Air Combat Command and the Air Mobility Command, provided important input to the committee. Finally, the original architects of the United States Navy's Naval Aviation Enterprise, including a past commander of Naval Aviation Forces and the current Director of Logistics for the Naval Air Forces, provided a valuable perspective on an alternative business model.[8,9,10] These latter Navy sources provided valuable insights and benchmarks on how the Navy operates its sustainment enterprise. Additionally, four committee members travelled to the Navy Fleet Readiness Center Southwest (FRCSW) to observe selected elements of the Navy's depot maintenance programs and the interfaces with program offices and supply support.[11] Although this chapter stands alone, topics such as policy, investments in general and in facility and equipment technology are covered in depth in Chapters 2, 3, and 5, respectively.

HISTORY AND PRESENT STATE OF AIR FORCE AIR LOGISTIC CENTERS

Historically, the ALCs have operated under several different organizational constructs. From the mid 1970s until approximately 1990, the ALCs had three primary functions, which, in today's terms, were material management, maintenance, and distribution. It is important to note that material management involved managing commodities and parts as well as weapon systems (i.e., system program offices for aircraft). There were also major support functions such as contracting and manufacturing, and communications and computer systems. Although the internal alignment of all of these functions varied during the 1970s to the early 1990s, the ALCs were consistently functionally aligned.

During the early 1990s the functions were re-aligned according to product orientation, which aligned weapon system sustainment not only with weapon system management but also with unique material management and major system maintenance, under a single leader. For example, at WR-ALC, a C-5 management directorate became responsible for C-5 aircraft systems, commodities, and maintenance. Also important is that the ALC commander, traditionally a major general, became responsible and accountable for the operation of the entire ALC organization. The commander generally had authority, responsibility, and accountability for the range of resources available within normal constraints to execute the programs

---

8Vice Admiral Walter Massenburg (United States Navy [USN], retired). "Enterprise Behavior: Fundamental Changes in the Government Business Model." Presentation to the committee, January 18, 2011.

9Vice Admiral James Zortman (USN, retired). "Aircraft Sustainment Strategy: Can Industry and Government Move the Needle?" Presentation to the committee, January 19, 2011.

10Captain Mike Kelly, Commander Naval Air Forces, Force Material. "COMNAVAIR. Presentation to the National Academies Committee on U.S. Air Force Sustainment." Presentation to the committee, January 17, 2011.

11A conscious decision was made to not visit United States Army sustainment activities. Unlike the Army, the Navy has significant commonality with Air Force systems and resourcing. Time constraints also did not permit such a visit.

assigned to the ALC. However, this no longer appears to be the case.[12] Since 1992, when Air Force Logistics Command and Air Force Systems Command were integrated into a single major command, the ALCs have been part of the AFMC. This consolidation placed ALCs, the Air Force Research Laboratory, and the Product Centers within the same major command. The AFMC develops, tests, fields, and sustains weapon systems. Today's ALCs provide some but not all weapon system and product support sustainment, some modernization, and some maintenance for the majority of the aircraft fleet. Some aircraft sustainment support is provided by the Product Centers, namely, the Aeronautical Systems Center and the Electronic Systems Center. The material management (supply chain) and distribution functions have largely been removed from the ALC commanders' responsibilities. At the time of the 1992 integration there were five major ALCs, and a series of changes began to chip away responsibility and authority from the ALC commanders. As a result of Defense Management Review Decision (DMRD) 902, the Air Force began to transfer to the Defense Logistics Agency (DLA) a sizeable portion of the ALC distribution functions, which essentially involved wholesale Class IX parts storage and the movement of parts around the ALC bases. The 1995 Base Realignment and Closure (BRAC) Act determined that the future Air Force depot maintenance work load could be more efficiently sized and managed in three ALCs. Consequently, the Sacramento ALC and the San Antonio ALC were selected for closure, with a large part of their maintenance and sustainment work realigned to the remaining three ALCs. Also in 1995, DMRD 926 began to transfer some non-recoverable parts (expendables) management to the DLA. Table 4-1 depicts the attributes and activities of the three ALCs.

With the 2005 BRAC, responsibility for greater portions of the ALC supply chain were re-aligned to the DLA. In addition, responsibility for managing nearly all remaining non-reparable items (expendables) was transferred from the Air Force to the DLA center in Richmond, Virginia. Although on initial examination the implications of any transfer of expendables seem rather routine, the fact is that these types of parts range in price from a few cents to tens of thousands of dollars per assembly (see Table 4-2). Importantly, these expendables are directly related to an ALC commander's maintenance production activities, and, whether pennies or thousands of dollars, when not available they can impact the commander's success.

The 2005 BRAC further eroded the ALC commanders' responsibilities with respect to two key aspects of supply chain management. First, the purchasing function for reparable parts at each of the ALCs was realigned to the DLA in place at the ALC base. In other words, although the Air Force would still manage reparable items, determine the quantities to be purchased, and initiate contracts for repair of these parts, the DLA would source and procure the new parts. The rationale for

---

[12]AFMC Historian Office. Personal communications to the committee, May 3, 2011.

TABLE 4-1 Air Force Air Logistics Centers and Their Attributes

ALC	Location	Total Base Employment	ALC Employment	MX Wing Employment
Warner-Robins	Robins AFB, Georgia	21,254	Total: 14,295; Civilians: 12,873; Military: 1,422	Total: 8,786; Civilians: 8,173; Military: 133
Ogden	Hill AFB, Utah	22,547	Civilian and Military: 13,483	7,796
Oklahoma City	Tinker AFB, Okla.	29,218	Total: 13,296; Civilians: 12,173; Military: 1,123	Total: 9,263; Civilians: 9,152; Military: 111

SOURCE: Air Force Materiel Command.

this transfer was based on the belief that consolidation would provide economies of scale and place major parts purchases under the management of the DLA—a single DOD supplies purchaser. The unintended consequence was to create numerous process seams and put the same part under two separate management systems, controls, and authorities. Under a March 2011 memorandum, the Secretary of Defense proposed a pilot program to move additional spare parts support to the DLA.[13]

[13]DoD. "Track Four Efficiency Initiatives Decisions." A Memorandum for USAF Key Personnel. March 11, 2011. Washington, D.C.: Office of the Under Secretary of Defense. Available at http://www.airforce-magazine.com/SiteCollectionDocuments/Reports/2011/March%202011/Day24/Secdef_Efficiencies_031411.pdf. Accessed May 3, 2011.

Maintenance Wing 2010 Revenue	Maintenance Floor Space	Weapon System and End-Item Program Management	Economic Impact
FY10 total revenue (including Material Support Division)—$1.738 billion	Maintenance Shops: 4,500 ft^2 Administration: 2,100 ft^2; Storage: 3,900 ft^2	E-8C, C-130, F-15, HH-60G, UH-1, U-2, DCGS, C-5, C-17, RQ-4, MQ-1, MQ-9, Software, Electronics, Electronic Warfare (missile/radar warning, chaff/flare dispensers, jamming), Support Equipment, Vehicles, Automated Test Systems, Commodities (C-5 flaps/pylons, C130 props/radomes, C-17 landing gear doors, F-15 wings/speed brakes), F-15 and C-130 FMS	$4.134 billion
$1.7 billion	17,400 ft^2	B-2 Structure, F-16, A-10, F-22, C-130, T-38, Commodities, Software	$3 billion
FY10 total; revenue (including Material Support Division)—$2.69 billion	9,200 ft^2	E-3, KC-135, B-1, B-52, KC-46A, B-2, Contract Logistics Support Commercial Derivative A/C (KC-10, E-4, VC-25, T-6, C-12, C-21, C-9, C-20, Peace Lotus, C-26, C-38, T-1A, Iraqi A/C, C-32/C-40, E-9, KDC-10, T-41, T-43, T-51, TG-10, TG-15, UV-18, C-37), Engines (F100, F101, F108, F118, F110, T56, TF39, TF34, TF33, F117, F119), Commodities, Software, ATCALS, HF Global, Foreign Military Sales	$3.511 billion

The 2005 BRAC's second change, equally significant in terms of impact, was the in-place transfer of Air Force personnel who directly provided parts movement and stocking to the ALC (almost 1,000 employees) to the DLA. These employees then reported to the DLA Distribution Command; they were no longer part of the organization responsible for producing the maintenance results, and their objectives and performance reviews were not driven by the ALC. Although almost 1,000 personnel were transferred, the Air Force has placed some personnel into positions to fill the voids in staffing and experience.

Following the 2005 BRAC, the Air Force established the Air Force Global Logistics Support Center (AFGLSC) to manage all reparable parts. This action occurred for a variety of reasons, including the desire to standardize processes for reparable

TABLE 4-2 Examples of Costs Relating to Spare Parts

Nomenclature	Quantity Used/Year	Last Purchase Price/ Repair Cost ($)	Extended Cost/Year ($)
Adj pin spacer tube (C-130)	22,913	7.71	176,659
Turbine rotor blade (F108 Engine)	Condemn: 806	6,375	5,138,250
	Repair: 5,511	287	1,581,657
			6,719,907

SOURCE: AFGLSC.

parts management across the three ALCs and to provide a standard approach to supply chain management for the external world to follow. The AFGLSC assumed responsibility for the Air Force reparable parts supply chain in March 2008[14] and is tasked with assuring the availability of reparable parts to both base-level maintenance activities and depot maintenance production activities. The current AFGLSC commander is a major general with personnel located at five or more locations within the Air Force.

The AFGLSC's creation completed the combination of changes that limited the ALC commanders' sustainment responsibilities to some weapon system or product program management as well as maintenance. The ALC commanders were rendered less effective because they lost responsibility, authority, and accountability for size\able portions of the supply chain, and had no real direct influence over outside suppliers but were still accountable for depot maintenance production and a portion of parts support to field-level activities.

Clearly, the ALCs have undergone a great deal of organizational change within the past 20 years. With this brief history in mind, it would be a mistake to view the ALCs in a monolithic fashion. The committee focused on the current state of the ALCs and their resourcing for present-day operations and future support of the assigned weapon systems. ALC leadership and authority, although not specific items in the TOR, are critical factors. For example, an ALC commander must deal with multiple lines of external authority, many of whom have far less experience and certainly far less of an Air Force enterprise view. An ALC commander should be considered a "supported commander," but many organizations fail to grasp the relationship of "supporting and supported" commander, and therefore operate independently and are driven by their own policies.[15] As a result, the ALC receives fragmented sustainment support, which impacts the support it provides to its own

[14]AFGLSC provisional command was established in April 2007.

[15]Kathy Cutler, Deputy Commander, Defense Logistics Agency Aviation. "Air Force Studies Board Committee on U.S. Air Force Sustainment." Presentation to the committee, February 17, 2011.

customers. The various entities, such as the supply chain and "out of the chain" command offices, optimize for their own organization and/or products rather than support a major sustainment entity such as an ALC. The various seams, the ALCs' limited ability to influence the supply chain, and the fragmented system alignment all impact outcomes—to the extent that it is surprising that the process works at all.

Finding 4-1. The ALC commanders' authority has been significantly weakened over the past several years to the extent that they do not have sufficient authority to effectively and efficiently execute the programs for which they are responsible.

Recommendation 4-1. The Air Force should establish streamlined command lines of accountability and authority to allow the ALC commanders clear execution authorities to direct process improvements on assigned programs, maintenance activities, and supply support.

HOLISTIC APPROACH TO THE RESPONSIBILITIES AND PERFORMANCE OF AIR FORCE AIR LOGISTICS CENTERS

It is essential to have a clear understanding of all the resources that impact the ALCs' operations to better evaluate the ALCs' capabilities. Today's ALCs have two principal focus areas: (1) weapon system or product management and (2) maintenance of assigned systems, end items, and parts. Numerous subsets of activities support accomplishments within these focus areas. For example, within weapon system and product management, there are product engineering, condition of equipment reviews, corrective action planning, modification management, maintenance planning and evaluations, configuration management, technical documentation updates, and safety processes. Likewise within maintenance, all actions are derived from engineering-based requirements with subsets that can and should lead to effective workflow. Among these are process engineering, workforce and workload planning and scheduling, replacement parts needs and bill of material upkeep, modern equipment technology insertion efforts, and facility management. All of these processes must come together for an ALC to fulfill its mission to provide support to the commanders who rely on the ALC products and services to fulfill their missions. To understand how these processes can come together, it is useful to review who and what roles are necessary to achieve the desired results.

To effectively carry out safe and efficient maintenance operations, a technical support process should be developed that begins with system design and ultimately extends to system retirement. From initial concepts, analyses are conducted to assess the system's purpose, mission, operating environments, and support exposure at all levels of the system or system operations. To be effective, these processes and analysis activities must form a continuum throughout the system's life cycle. To this end, the ALCs play a key, but not the only, role. The ALC role is part of a complex process involving the system users (operating commands), program management offices, contractors who manufacture and provide all types of support, and Air Force laboratories that assist with technology infusion as well as with modernizing equipment for system maintenance. Effective and efficient sustainment programs are directly tied to the coordination of these multiple players. Achieving effective and efficient sustainment programs is a direct responsibility of many, but should be an enterprise integration effort of the ALC commander.

Sustainment activities should be focused to maintain a system in the status for which it was designed. Thus, the activities should interface with all aspects of operations and processes to ensure that the design goals and or mission objectives are accomplished. The activities should involve a series of tasks that can maintain operations in an orderly fashion throughout the system's life cycle. The purpose is to monitor the system's condition to assure that it does not inhibit the design goals or mission effectiveness. An example of such an effort would be strong data capture and archiving. How data processes enhance this support is covered in detail in Chapters 2 and 6.

Program management can substantially improve the overall operation of a product but must take a holistic approach to do so. Maintenance alone cannot make something better than the design objectives without integrated support from the program management engineering. In the absence of desired improvements to the design, maintenance can only assure that the system does not deteriorate below the design objective. Maintenance must be supported by all the factors of program management and provided with what it needs, when it needs it, by the entire supply chain. Consequently, the committee viewed sustainment activities and the resourcing of these activities at the ALCs in a holistic manner. During meetings with sustainment officials, the discussion of support for the ALCs centered on policy, planning, people, products, processes, and parts. In addition, the organizational structure, technology infusion, and plant equipment were assessed. It is important to realize that all of these factors are inter-related and inter-dependent on an ongoing continuum that affects the ability of the ALC to meet the sustainment goals that are established above the ALC commander.[16]

In measuring the resourcing of the ALCs—funding, workforce, skill sets, technologies—the committee studied a large number of charts and various planning documents, including the ALCs' strategic plans. These plans significantly differ from ALC to ALC. Aircraft availability (AA) was often mentioned as the principal measure of merit for the Air Force's as well as the ALCs' sustainment goals. However, after analyzing AA, the committee found many external factors that affect

[16]See Chapter 2 for a discussion of the Air Force's sustainment goals.

AA and that are well beyond the ALC commander's ability to control. As discussed in this chapter and in Chapter 2, the committee was unable to identify succinct sustainment goals that the Air Force can use to measure the ALCs' success. Later in this chapter the report discusses metrics on aircraft production that were used at the United States Navy's FRCSW and that captured a best practice on aircraft production quite well.

> **Finding 4-2.** The metrics that would determine the success of an ALC, such as cost, schedule, and performance, are not widely used and consistently applied across the three ALCs.

As the following section shows, the broad subject of resourcing impacts the ability of the ALCs to accomplish their missions now and in the future.

ASSESSMENT OF THE RESOURCING OF AIR LOGISTICS CENTERS

Policy-Driven Supply Support

Policy for sustainment is largely covered in Chapter 2. As the committee assessed the resourcing of the ALCs, it found a great deal of policy that can, in various ways and through various interpretations and regulations, encumber the efficiency of the processes. Policy is always essential; however, the Air Force's existing policy does not clearly identify the official responsible for ALC sustainment activities and whether that official has the proper resources to execute the mission. The committee repeatedly asked the questions: "Who decides what gets executed at ALCs?" and "Who determines what resources the commander has to execute the mission?" The answers were usually nebulous and lacked specificity but were mostly along the lines of "It depends." *There is clearly a major disconnect between the policy's intent and its execution.* To build on the earlier discussion of supply chain fragmentation, the following discussions offer a few specific examples of organizations external to the ALCs' executing decisions that affect the ALCs' efforts.

At WR-ALC, the C-17 aircraft are modified and repaired by government employees. Spare parts are supplied by contractors, and even though this process performs remarkably well, it has been proposed that spare parts be supplied by organic Air Force and DLA sources.[17] In discussing this concept with C-17 main-

[17]During briefings at WR-ALC, the source of supply was a subject of great concern. During subsequent discussions, it became evident that a final decision on the C-17 supply support has not been made. However, it is noteworthy that the widespread conviction of government maintenance personnel is that the contractor-operated supply chains provide far better support than government-operated supply chains.

tenance managers and technicians, the committee learned that a contractor, when responsible for a part, can typically supply the part within 24 hours, while organic sources routinely take 17 days. Although it is expected that a contractor will take more time as the C-17 production activity draws down, it still will not take as much time as the government currently takes. Additionally, the committee repeatedly heard comments about poor supply support to repair lines across all three ALCs. The C-17 proposal to move to organic support and similar stories on other product lines are key examples of how the ALCs are not being resourced to meet production demands. When parts support across all ALCs is considered, the poor performance is systemic, and sustainment activities are not resourced for effective and efficient operations.

Resourcing the ALC Workforce

The most critical resource for the ALCs is a workforce that is remarkably talented and goes to extremes to make a complicated organizational environment work. The committee interfaced with employees in conference room settings and by "walking the floor" to talk with them at random. Employees at all levels exhibited great attitudes and were frequently well-versed about, and clearly committed to, improvement activities. Safety practices and a keen awareness for environmentally "green" operations were observed at each ALC. Overall, the general industrial operations as well as the product support sustainment workforces are producing reasonably well. There is a willingness to change to improve processes, but pockets of insular thinking remain, which is not unusual in organizations the size of the ALCs. The leadership is clearly committed to improvement; however, they do not have command-standardized processes, metrics, and goals.

To the extent possible, the committee analyzed workforce levels to support the ALC sustainment missions. The ALCs have reasonable distributions of staff with skill sets to meet current needs. However, the ALC workforce appears to be constrained in product support sustainment activities in much the same way that other similar Air Force-wide organizations are suffering workforce shortages. The workforce may be further constrained because a definable workforce standard does not seem to exist for the program offices engaged in sustainment or for the AFGLSC. The committee could not determine what levels of staffing are sufficient or required, because the amount of predicted work varies, and there is no definite operating plan to support staffing levels.

For industrial activities, the workforce seems to be more balanced to the workload. Yet, the ability to rapidly adjust staffing levels, as would happen in the commercial sector as workload changes, is constrained in the ALCs. Despite a known greater than 10 percent growth in maintenance workload, an ALC was restricted by higher levels from hiring maintenance personnel to meet the growing workload. In fact, the ability to hire as workload grows was constrained for some time, and this direct employment "cap" has contributed to higher-than-desired work-in-process and year-end carryover. An analysis of the Air Force industrial workload personnel policies indicates that greater freedom to adjust staffing levels without higher Headquarters interference is needed. [18] This is not to say that the process should be free from executive reviews. However, ALC commanders should be able to rapidly adjust staffing levels to effectively and efficiently meet fluctuating workload demands. The net result of the delay was a significant growth in work-in-process, unfavorable year-end carryover, and the attendant budget impacts. However, the more significant result was that aircraft became backlogged in the depot maintenance process, and there were fewer aircraft available to the operational commands. What appears to be a budget-driven decision on workforce management may have netted short-term benefits, but its long-term impacts on the Air Force and ALC work management were very disruptive. The occurrence of such actions was surprising, but not as surprising as the long-term outcome. In the commercial sector, flexibility exists to flow workforce with workload. The commander and the civilian executive director at the FRCSW both commented that, although their industrial workforce levels are reviewed, they have freedom to adjust industrial personnel levels as determined by the workload.[19]

50/50 Affects the ALC Workforce

Another policy issue with broad ramifications as well as a direct impact on the workforce, which was discussed in depth in Chapter 2, is the requirement to have at least 50 percent of the depot maintenance performed in government facilities. This is commonly referred to as the 50/50 rule of 10 U.S.C. § 2466. The genesis and purpose of the rule are understood; however, its constraints impact product support, depot maintenance planning, and warfighter asset availability. At the ALCs, the impacts are felt on the production lines, where work-in-process increases as the Air Force directs earlier than planned maintenance to assure that the 50/50 rule is not violated during a fiscal year. Although it is recognized that some attempts have been made to change what is or is not in the 50/50 equation, these attempts have not been successful to date.

The effects of surged workload on the workforce are obvious. But the problem is not the 50/50 rule—it is the Air Force's fragmented approach to managing 50/50

[18] General Military Law, U.S. Code, Title 10, Section 2472, February 10, 1996.

[19] Captain Fred Melnick, USN Commanding Officer, and Mr. William Reschke, Plant General Manager, USN, Fleet Readiness Center Southwest (FRCSW), San Diego, California, personal communications with the committee, March 29, 2011.

on a long-term basis.[20] The Air Force should undertake aggressive, positive, long-range planning on a large scale to assure that in the near future the 50/50 balance is sufficient and the ratio is not threatening. To not do so ignores the fact that Title 10 has put the components on notice to achieve, and not necessarily cause, significant perturbations in the workforce.

Critical Support Staff for the ALC Workforce

Support manpower issues affect the ALC workforce. These issues directly pertain to the support forces that allow the depot maintenance production effort to proceed efficiently and effectively and have been referred to as "color of money" issues. However, the issue actually centers around proper sizing of the support workforce—a challenge that extends into the working level and impacts support to the production lines. As an example, when the workload for a product line increases, the workforce will eventually increase to meet the workload, but the engineering support that provides technical direction and instructions for the craftsmen comes from a different pool of authorizations and resources. Thus, the production workforce and engineering support do not uniformly grow to meet the increased demand. As systems age, the situation becomes increasingly difficult to manage because the workload and the demand for engineering support usually increase commensurately.

A second area, like engineering support, is DLA support to the production lines. Although the workload for selected repair lines has increased, the size of the DLA workforce that supports parts to these repair operations has not been linked to the repair volume. DLA Aviation reported that the DLA employee count is linked to government-wide personnel policies.[21] The hiring restriction may be temporary, but the fact will remain that the DLA manages personnel independent of the repair effort, which was confirmed in discussions at the ALCs and with DLA Aviation. Thus, while the workload on repair lines changes, portions of the workforce that support parts for repairs lines are completely disconnected from the function they support. These are just two of a select few of the many policy impediments that affect the ALCs in their sustainment efforts.

Finding 4-3. There remain significant issues related to providing a skilled product engineering and DLA supply support workforce to ALC maintenance activities to assure that the support is properly sized to match the workload.

[20] Recommendation 2-2 in Chapter 2 provides a related recommendation.

[21] Kathy Cutler, Deputy Commander, Defense Logistics Agency Aviation. "DLA Aviation." Presentation to the committee, February 17, 2011.

Workforce Training

The ALCs have established notable industrial-type training for new employees. ALCs have close relationships with local area community colleges and technical schools for training of new industrial employees as well as "upgrade" training where appropriate. These training programs have allowed the ALCs to fairly rapidly meet significant increased demand for shop floor personnel due to retirements and work scope increases.

While technical skill training is ongoing, the significant growth in shop floor personnel has generated a requirement for training incoming workers on the detailed repair and overhaul procedures unique to specific work stations. Many of these techniques were not reduced to writing or a knowledge capture basis and consequently require a "revised" learning curve. Improvements in knowledge capture from vested employees are being addressed with the retiring workers, whose experience is being documented for all critical processes. The introduction of low observable technology presents additional workforce training requirements that must be addressed in the future.

The ALCs recognize that strong leadership and supervisory skills "don't just happen"—they develop over time and can only come from efforts by leadership. Figure 4-5 depicts the "tip of the iceberg" of what training programs can provide. It is most noteworthy that 1,300 applicants sought this type of professional and personal development in its first year alone.

Labor Relations and Growing the Workforce

A key part of the ALCs' ability to perform effectively is their relationship with the unions. There appeared to be a distinct and noticeable difference in the relationship between ALC management and unions at the various ALCs. The ALC leadership teams were strongly committed to partnering with the unions but met with varying degrees of success. There is room for significant improvement in gaining flexible work rules to permit the workforce to be more versatile. Such a change would improve utilization of available resources, allow adaptation of improved processes, and gain efficiencies and productivity. The Air Force should recognize that relief from overly restrictive policies, which in a modern industrial aviation setting impede effective and efficient production.

Finding 4-4. The ALC workforce is professional and continues to work within the Air Force system to provide strong weapon system sustainment. The leadership teams are engaged with the workforce and constantly demand process improvements and efficiencies; however, the current work rules lack the flexibility to achieve these efficiencies.

FIGURE 4-5
Supervisor development program at OC-ALC. SOURCE: Major General P. David Gillett, Jr., Commander, OC-ALC. "OC-ALC Strategic Goals." Presentation to the committee, January 11, 2011.

Recommendation 4-2. The Air Force should follow in a more timely manner the statutes that require the depot maintenance industrial workforce to be managed according to workload. The Air Force should also ensure that supporting organizations are staffed to support the industrial workloads and that flexible work rules are established to permit more workforce versatility.

ORGANIZING THE AIR LOGISTIC CENTERS FOR SUSTAINMENT

Organizational structure and process flow is an important part of the resourcing equation. The ALCs support the "supported commander." Yet, success is often attributed to the individuals involved in "making it happen" rather than to an optimal, high-efficiency construct, clearly defined lines of authority, effective enterprise resource planning systems, and data-driven analysis and actions. For example, programs differ as to the roles and responsibilities of the program manager, sustainment manager, system support manager, systems development manager, product support manager, and a Program Executive Officer.[22] The committee was often told of the frustration in executing programs and controlling resources. Fig-

[22]The committee heard all of these terms used, often interchangeably.

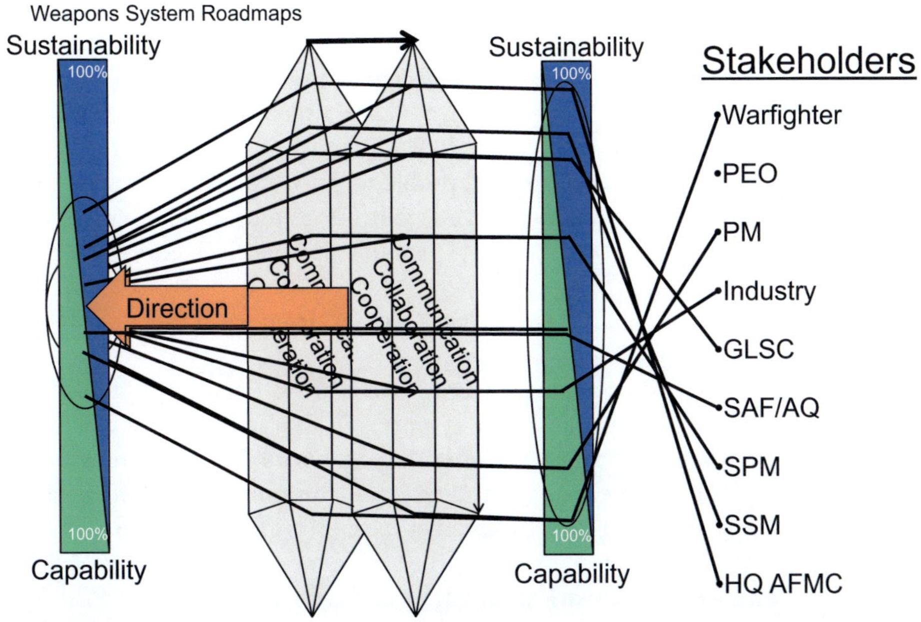

FIGURE 4-6
Focus the enterprise. SOURCE: Brigadier General Arnold W. Bunch, Jr., Director and Program Executive Officer for the Fighters and Bombers Directorate, Aeronautical Systems Center, Wright-Patterson Air Force Base. "ASC/WW (Fighter/Bomber PEO)." Presentation to the committee, December 8, 2011.

ure 4-6 illustrates the lines of authority and communication among the various Air Force sustainment stakeholders.

When this information is reviewed and considered in the context of the major operations at the ALCs, it becomes apparent that the organizational structure is so dispersed that no one is responsible. One can only imagine how this diagram would look if expanded to represent all of the systems that are supported by the ALCs. The lines of authority for multiple programs become so convoluted that personalities prevail rather than defined processes achieving optimal results. Without clearly defined responsibilities and prudent metrics to determine results, system breakdowns are commonplace. Unfortunately, this allows for excuses and often leads to celebrating superficial success rather than measurable real achievements.

As discussed earlier in this chapter, the spare parts organizational structures for the Air Force at large and for the depot maintenance lines, in particular, are broken, largely because no one officer is responsible for the supply chain. The committee repeatedly heard that expensive hanger space, tooling, equipment, and, most importantly, skilled technical personnel are either forced to conduct workarounds, cannibalize parts, or place items in a downtime status for the lack of parts. *These events are not exceptional; they occur with sufficient frequency that shop floor employees expect them, take them for granted, and freely shared their frustrations about them with committee members.*[23] Although the DLA is often cited as the largest offender for parts shortages, the Air Force also has problems providing parts. The Air Force stated that the 20 percent unavailability is partially due to the unavailability of DLA parts needed to support the repair of Air Force managed parts. Figure 4-7 depicts the spare parts situation.

In some cases, the parts shortages reflect poor performance of the supply chain, and in others the Bills of Material and the quantity of items to be repaired were not properly defined to the supplying agencies. From a broad perspective, the overall parts situation is just another reflection of who is in charge (or not in charge) of the sustainment process at the ALCs. Under the organizational structure, the ALC commander runs the facility as a coordinating and communicating officer rather than an executive with authority to direct all of the key elements of the sustainment process. As the Air Force moves into new production modes such as High Velocity Maintenance and Repair Network Integration that demand higher flows through the repair cycle, the current parts situation will not be capable of sustaining these new processes. Dramatic improvements are needed.

As another classic example of organizational dysfunction, nearly 2 years after the DLA assumed responsibility at one ALC for managing the stocking and the parts movements to the shop floors, debate continued over who was responsible for providing kitted parts to the mechanic level. At another ALC, the DLA was "kitting away," but it wasn't clear who was building the kits. In either case, clear lines of authority do not exist to resolve disputes and expedite decisions on supply chain issues.

The DLA and the Air Force have established metrics on the performance of the DLA enterprise to the Air Force.[24] Likewise, the senior DLA official at OO-ALC provided metrics on parts support to that site. A Performance-Based Agreement (PBA) has been executed between the Air Force Deputy Chief of Staff for Instal-

[23]In fact, in the Air Force, complex parts are routinely not available within 2 days 20 percent of the time. The less complex parts provided by the DLA are not available 10 percent of the time. The committee was consistently and independently told that the maintenance leadership teams must turn to four to six supply chain managers/organizations to find a responsible person who might be able to resolve the spare parts problem.

[24]Kathy Cutler, Deputy Commander, Defense Logistics Agency Aviation. "DLA Aviation." Presentation to the committee on February 17, 2011.

lations, Mission Support and Logistics and the Director of the DLA.[25,26] Close examination of the document reveals that nine key metrics have been agreed upon for use today. These metrics represent the first concrete evidence that performance of the supply chain is being measured. The concept of supply chain metrics should be continued; however, the nine metrics in the PBA do not adequately reflect the conditions of the supply chain.

Foremost, the key metrics measure degrees of improvement that are modest at best. The majority of the metrics have no upper or lower threshold limits, and the others identify only degrees of improvement instead of a standard to be achieved. Although continuous improvement is important, when resources are finite and demands great, metrics without standards to be achieved are confusing to the responsible parties, especially. That is, how do they know when they have achieved the desired performance level? It is like a basketball team high-fiving because they narrowed the points margin, but they are still down 20 with a minute left in the game! One metric of the nine, Mission Incapable Awaiting Parts (MICAP), generally measures "the pain" of not having parts and applies to both field and depot operations. However, the target goal for FY2011 is expressed in a table for each weapon system/subsystem and does not begin to portray the degree of parts non-availability for selected product lines at the depots, in part because depot maintenance activities can express MICAP conditions based on work content, timing, and product lines. This PBA is a good first attempt to measure the DLA performance, but additional effort should be devoted to developing metrics with upper and lower thresholds, thresholds and standards as appropriate, and a key single metric that reflects the impact on operational effectiveness and most importantly on efficiency of field and depot maintenance operations of the DLA-managed supply chain.

A presentation at a recent symposium outlined the complexities of the marketplace for any entity that deals with external suppliers.[27] Figure 4-8 outlines a strategic framework that challenges supply chain managers. As the criticality and complexity of the items to be supplied increases, the options for the procurer become more limited, and the options for the supplier become more customer-focused than commodity-focused.

Ultimately, the supplier- buyer relationship must mature into a partnership arrangement. As Figure 4-9 shows, with the more complex parts used in the aviation world today, product specialty suppliers such as original equipment manufacturers

---

[25]Ibid.

[26]USAF. 2010. Logistics, Installations & Mission Support United States Air Force (AF/A4/7) and Defense Logistics Agency (DLA) - PERFORMANCE BASED AGREEMENT (PBA), Version 3.0. Sept 20. Available at https://acc.dau.mil/CommunityBrowser.aspx?id=32525. Accessed March 1, 2011.

[27]Steve Geary, Center for Executive Education, University of Tennessee. "Acquiring What the Warfighter Needs." Presentation to the Aviation Week MRO Conference, April 12-14, 2011, Miami, Florida.

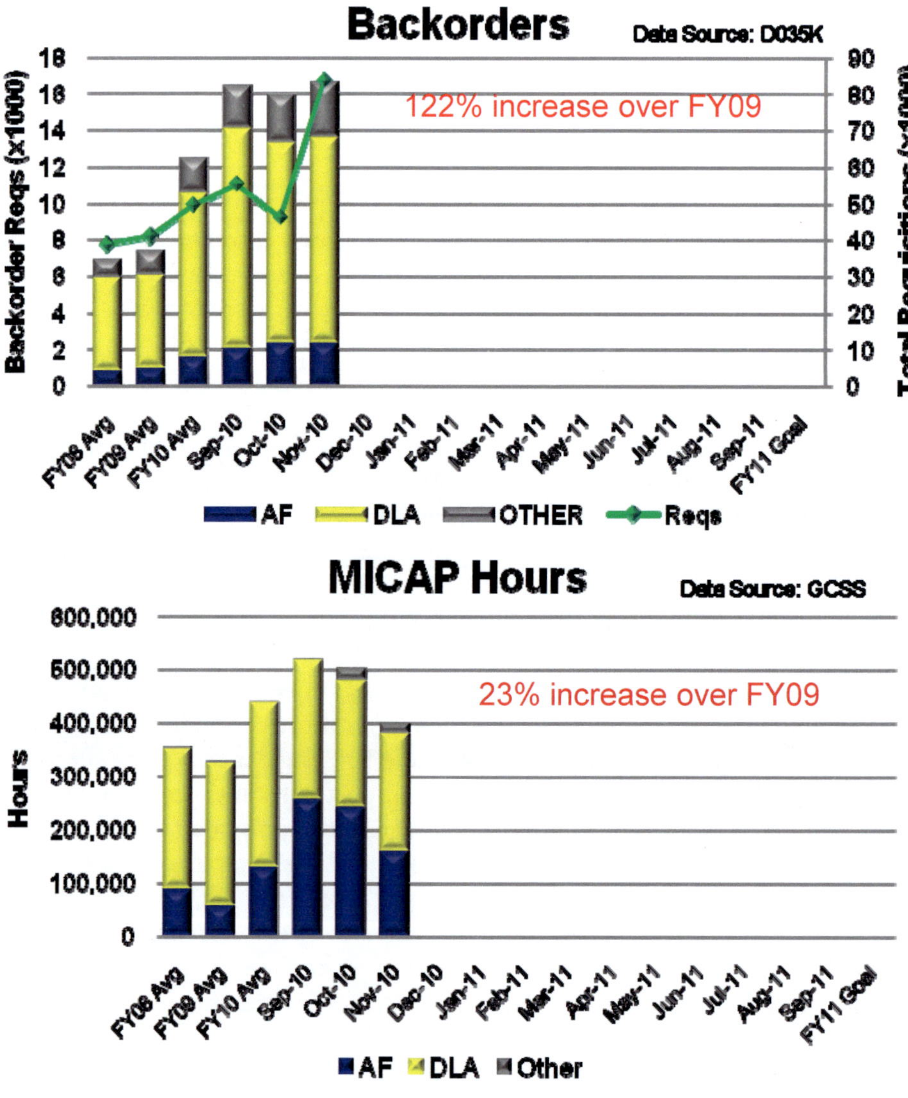

FIGURE 4-7
Supply support to 76 MXW by source of supply. MICAP, mission incapable awaiting parts; PDM, programmed depot maintenance. SOURCE: Major General Bruce A. Litchfield, Commander, 76 MXW, OC-ALC. "76 MXW Production Machine." Presentation to the committee, January 11, 2011.

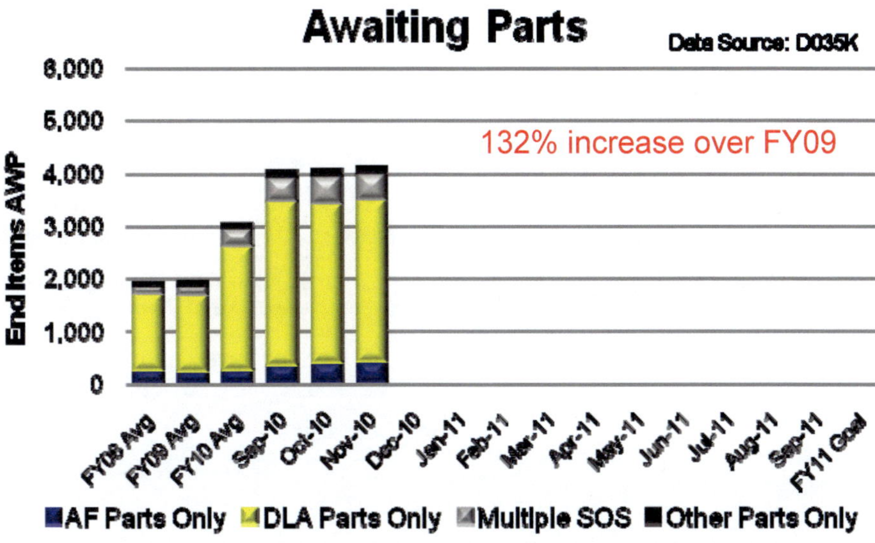

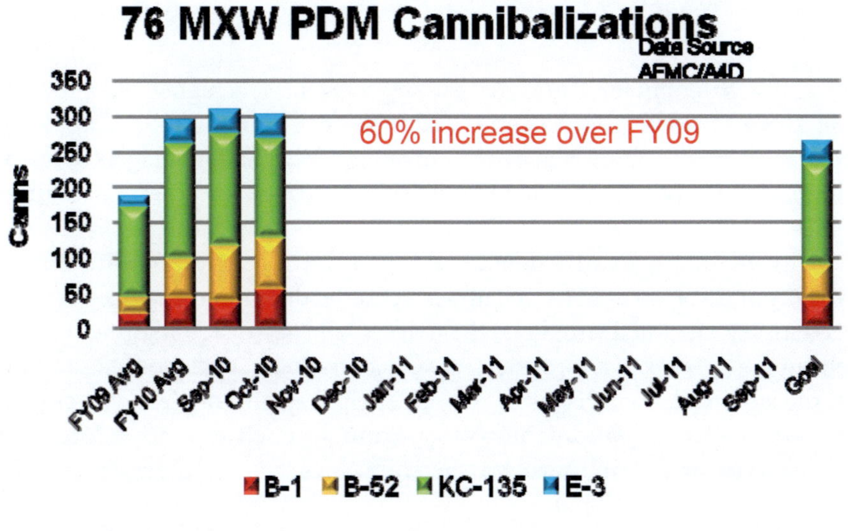

FIGURE 4-7 *continued*

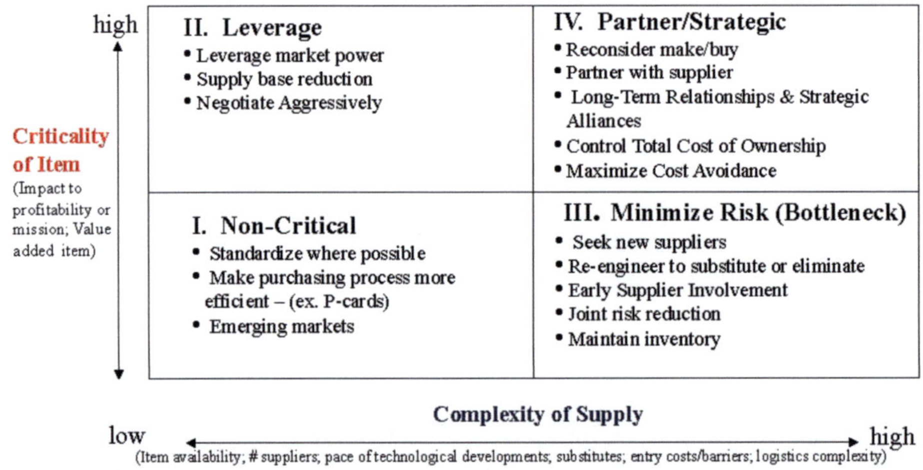

FIGURE 4-8
A strategic framework. SOURCE. Peter Kraljic. 1983. "Purchasing Must Become Supply Management." *Harvard Business Review* 83509 (September-October). Available at http://www.sourcingchina.org/PictureLoad/2009829103938430.pdf. Accessed May 5, 2011.

(OEMs) and high-confidence aftermarket suppliers have a higher likelihood of providing the parts where and when needed than do mass commodity suppliers.

The present market approach by the current supply chain managers seems to drive all classes of supply to the characteristics listed in the lower left quadrant of Figure 4-10.

In reality, market dynamics demonstrate that for many commodities this is entirely appropriate. However, for the upper tier of readiness and engineering critical parts with few qualified suppliers, the market approach with alignment to key qualified suppliers is more appropriate. A subject matter expert would also point out that the supply characteristics in the upper right quadrant (few participants, high barriers to entry, high switching costs) cannot be left to market forces but must be cultivated with long- term partnerships. Figure 4-11 highlights partnerships strategies to be considered.

These partnerships must be strategic and tactical, and both sides have an obligation to control the effectiveness and efficiency outcome. *Establishing long-term spare parts relationships with suppliers who are capable of providing parts and particularly those that fit in the upper right quadrant of Figure 4-9 would provide highly effective (minimum lead time, on-time delivery) and optimally efficient results (appropriate costs).*[28]

[28]Peter Kraljic. 1983. Purchasing Must Become Supply Management. Harvard Business Review 83509 (September-October). Available at http://www.sourcingchina.org/PictureLoad/2009829103938430.pdf. Accessed May 5, 2011.

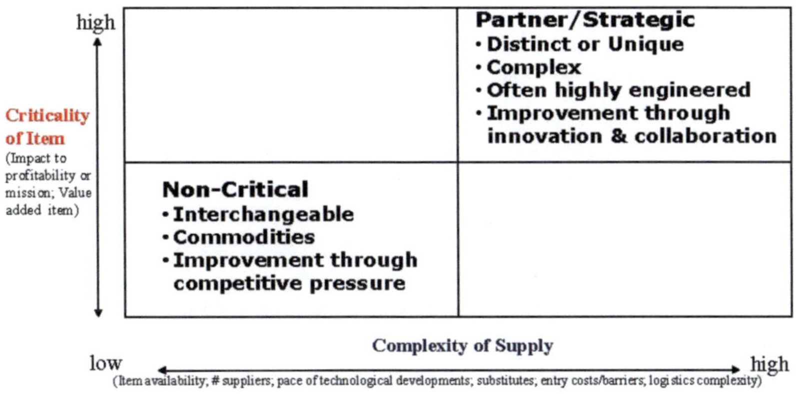

FIGURE 4-9
Production characteristics. SOURCE: Steve Geary, Center for Executive Education, University of Tennessee. "Acquiring What the Warfighter Needs." Presentation to the Aviation Week MRO Conference, April 12-14, 2011, Miami, Florida.

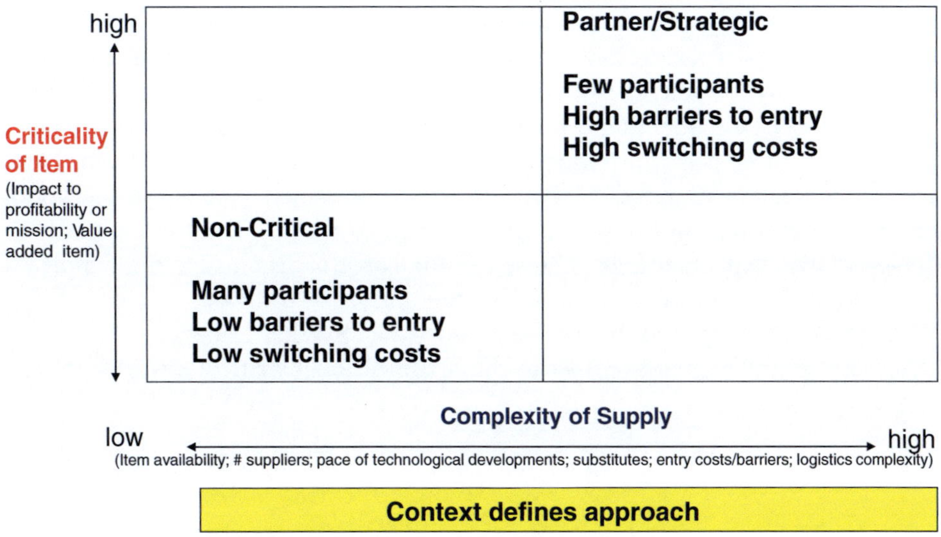

FIGURE 4-10
Production characteristics. SOURCE: Steve Geary, Center for Executive Education, University of Tennessee. "Acquiring What the Warfighter Needs." Presentation to the Aviation Week MRO Conference, April 12-14, 2011, Miami, Florida.

What strategies can we consider?
Some ideas based on observations

- Ensure properly structured long-term relationships to promote competitive pressure in monopsony.
- Gainsharing helps get everybody on the same side.
- Quantify the cost of achieving the cost reduction by tying required investment to the length of the commitment.
- Explicitly agree on the scope, the approach, and the targeted costs. Make sure they tie to the budget.
- Model the end-to-end costs. Finding ways to make fixed costs variable creates spend flexibility and reduces friction.

FIGURE 4-11
Production characteristics. SOURCE: Steve Geary, Center for Executive Education, University of Tennessee. "Acquiring What the Warfighter Needs." Presentation to the Aviation Week MRO Conference, April 12-14, 2011, Miami, Florida.

Finally, the DLA and the AFGLSC each have a colonel designate as the key coordinator of spare parts support at the respective ALC. The AFGLSC colonel is a commander and seems to have extensive decision-making authority for the personnel and parts execution of his group. The DLA colonel serves as a point of contact with whom ALC personnel can work. Yet, the DLA colonel must defer to lower levels at DLA Richmond for decisions such as local manufacture or rapid sourcing to meet demands, which are significant to the production effort but inconsequential in terms of difficulty or economic value in the bigger picture of enterprise resource consumption. In a briefing to the full committee, the DLA representative explained that because the Air Force is responsible for significant portions of the approval flow, DLA Richmond must be involved.

During discussions at the ALCs and at the FRCSW, key leaders made very appropriate comments with respect to the organizational structure, workforce,[29,30]

[29]Major General Andrew E. Busch, Commander, OO-ALC. Personal communications with committee, January 31, 2011.

[30]In general, Major General Busch stated that the issues affecting performance are not so much about organizational structure as they are about defining and following processes. In other words, organizational change will not necessarily solve the many issues, but crafting and following defined process may.

and spare support.[31,32] At the Aviation Week MRO conference, the Assistant Deputy Undersecretary of Defense for Maintenance succinctly summarized the bottom line for any maintenance effort: "If you don't have parts, you can't do maintenance."[33]

Finding 4-5. The current supply chain for spare parts, as managed by the Air Force and the DLA, is deficient and often fails to provide parts for technicians in time to affect repairs in an efficient manner.

Recommendation 4-3. The Air Force should take action, both internally (with Air Force sources of supply and Air Force-controlled contractors) and externally across the DoD (DLA), to dramatically improve the supply chain's timeliness and accuracy in providing necessary parts to meet the maintenance technician's needs.

RESOURCING FOR TECHNOLOGY INSERTION AT AIR LOGISTICS CENTERS

Although technology insertion is covered in depth in Chapters 5 and 6, the committee notes here that it observed a limited amount of technology insertion into weapon system and commodity parts and into supporting maintenance equipment at the ALCs. Insertion was largely a function of what was available and provided to program management as funded by operational commands. The committee did not detect any reluctance or inability by ALC program office personnel to accomplish technology insertion, including the associated master planning and the execution, when opportunities arose. Improving the levels of the manufacturing technology (ManTech) program and Component Improvement Program (CIP) appears to be a problem within the broader AFMC organizational structure rather than at the ALC level. At the ALCs, a viable organizational structure and good support exist to institutionalize the process of developing and transitioning repair and durability. Repair technologies—that is, the capability to repair systems and parts—at the ALCs appear to lag behind the introduction of new systems, in part because design data packages are lacking early in the fielding of new systems and because the ALCs initially rely on contractor logistics support, which is a significant chal-

[31]Captain Fred Melnick, Commander, United States Navy, Fleet Readiness Center Southwest. Personal communication, March 29, 2011.

[32]Captain Melnick commented on the DLA's metrics in the sense of the Air Force's PBA with DLA. In general he stated, "The metrics may mean a lot to DLA, but they do not have the same meaning to 'us.'" That is, the metrics do not measure the true supply support to the field-level activities that are receiving supply support from the DLA.

[33]John Johns, Assistant Secretary of Defense for Maintenance. Presentation at the Aviation Week MRO Conference, April 12, 2011, Miami, Florida.

lenge to meeting the 50/50 rule. Recent policy decisions to at least source and price data packages are noteworthy endeavors.[34] The resource impacts of procuring data packages are understood, and it is encouraging that a more methodical approach to tie data packages to the analyses conducted in support of DoD Instruction 5000.02 Milestone A and Milestone B activities has been observed. This effort appears to be progressive for the longer term, but in the immediate and near term the ALCs are impacted by lack of strong data packages. The overall data situation is made worse by efforts to manage software that is entirely dependent on total data packages being available. In the end, data are key to the ALC business of sustaining systems.

ENTERPRISE MANAGEMENT AT THE AIR LOGISTICS CENTERS

For the ALCs to be effective, data must become the key element in their efforts.[35] Chapter 2 addresses how data suppots engineering. Maintenance of complex equipment is as much about information as it is about the ability to execute the actual work. The lack of effective measures, an effective planning and forecasting system, true condition monitoring of the equipment, and enterprise capture of work accomplished are major impediments to the ALCs' success. During the assessment of the ALCs, much was heard about the Expeditionary Combat Support System (ECSS) and the promise it holds for the future sustainment processes management. Although enterprise management extends far beyond the ALCs, it certainly is a substantial part of the ALC environment and is a key part of the ALC resourcing. ECSS is a component—albeit an important component—of the Air Force Expeditionary Logistics for the 21st Century (eLog21) initiative. Numerous processes under eLog21 are designed to improve the governance of sustainment and the effectiveness and efficiency of the processes.[36]

ECSS standardizes sustainment processes at every level and uses an Enterprise Resource Planning (ERP) tool to enforce the approved rule set. Although this is noteworthy, the committee also detected the "pain" of legacy systems that have been forced to operate "as is" for years while ECSS has been fielded. Several committee members have fielded ERPs and agree with the decisions to "freeze" the legacy systems. This approach creates a very high risk for the near term because there are no other systems to take their place. Thus freezing of legacy system improvements requires considerable faith in those responsible for current mission execution that a very large and complex system will in fact be implemented in a timely fashion.

[34]DoD. "Better Buying Power: Guidance for Obtaining Greater Efficiency and Productivity in Defense Spending." A Memorandum for Acquisition Professionals. September 14, 2010. Washington, D.C.: Office of the Under Secretary of Defense. Available at http://www.acq.osd.mil/docs/USD_ATL_Guidance_Memo_September_14_2010_FINAL.PDF. Accessed May 5, 2011.

[35]Further discussion of data and enterprise management occurs in Chapters 2 and 5.

[36]eLog21 is discussed extensively in Chapter 2.

Based on experience, the skepticism in various execution offices is well founded. The legacy systems with all of their shortcomings are the only systems available. As a result, today there is no integrated or enterprise system that ties all of the sustainment requirements together, easily and accurately ties maintenance consumption to forecasts for future production needs, allows for the engineering officer to manage the mechanic information flow, tracks man-hours consumption or task completion, assists with planning and scheduling, or allows supply systems to meet the consumer demands. The lack of an integrated planning tool is a serious resource constraint for the ALCs.

During tours at one repair facility, the committee witnessed a locally developed "online" report that used legacy systems to draw information and allowed the maintenance personnel to see worldwide demands for the products, to track month-to-date production against a month-to-date requirement, and track current critical product needs. The committee asked maintenance leadership, "What other production activities has this system been exported to?" The response was noteworthy in that it simply hadn't been exported, for reasons related to the promise of a future with ECSS, a need to tailor the report somewhat to meet other area's needs, a reluctance by some to adopt the system in light of the coming of ECSS, and limited funding to alter the system to work in other areas. All of the reasons demonstrate the strong need for a standard way of doing business. The report that the committee observed, while very good, is one of a kind and stands alone in this particular repair group. In one form or another, the committee repeatedly heard similar stories about the condition of management systems to support parts requirements generation or production support. Suffice it to say, the sustainment process is hindered by the currently fielded or lack of process management and reporting tools.

A program such as eLog21 and its ECSS component are very valuable and are a necessary way forward. Further, these systems are by their nature expensive, have an element of risk, and take time. In pursuing an ERP system, whether it is ECSS or others, the Air Force must understand that implementation of major systems requires extensive buy-in from top to bottom, stable management, and process and work habit changes that are very disruptive. In such a major system implementation, frequent organizational changes are anathema and will greatly hinder the success. Major system changes cannot be implemented by outside consultants or contractors alone but must have major engagement of the actual management and workforce that will use them. Effective ERP systems are the cornerstone of an effective sustainment effort. Over the course of the study, the committee reached the conclusion that one of the single greatest impediments to a successful sustainment enterprise is the lack of standardized processes and a functioning enterprise-level data system that forces standardized actions and delivers timely, actionable data to all levels of work/management.

Finding 4-6. The lack of a functioning ERP tool or a strong management information suite and the reliance on outdated legacy information systems are great causes of added complexity and inefficiency in the operation of the entire sustainment enterprise that includes system "cradle to grave" program management, and maintenance, supply, and transportation activities.

Recommendation 4-4. The Air Force should continue its eLog 21 approach to sustainment improvement and should aggressively continue to pursue incremental fielding of the ECSS as an enterprise resource planning solution. Strong advocacy for this program should reside in the sustainment commander.[37]

MAINTENANCE PLANNING AND PRIORITIES

During tours of the ALCs, the committee searched for evidence of strong planning and execution of maintenance programs. Improvement programs for existing system are being planned, such as High Velocity Maintenance (HVM) and Maintenance Steering Group Three. These efforts are designed to enhance depot throughput and to tailor both depot and field inspections to realistic operational conditions. Likewise the committee saw evidence of lean practices and critical chain project management tools that will improve the overall maintenance processes. The full-scale deployment of these improvement programs is not envisioned, and programs such as HVM have different meanings for WR-ALC, OC-ALC, and OO-ALC. Often these programs are local efforts with little or no central focus on efficiency improvements. Although it was referenced by Headquarters staff, the committee did not see a concerted effort on the part of the AFMC to implement and measure productivity measures systemwide. Until recently, HVM was in prototype stage. The new WR-ALC commander instructed the HVM team to proceed and plan on full implementation.[38]

The maintenance staffing on aircraft during heavy maintenance is certainly lower than that experienced in the commercial aviation industry. The committee notes the new programs and use of enhanced productivity efforts; however, appropriate man loading would complement all of these efforts, more quickly improve flow through the depot, reduce the work-in-process at the installation at any one time, and still produce the same output per year. The committee was told that increased man loading cannot be done because the ALCs have not invested in a

[37]Based on detailed information provided to the committee by the Air Force, the committee concluded that there is a critical need for an ERP system to help address Air Force sustainment issues. The ECSS is being developed with ERP capabilities and will be implemented to meet the Air Force need. The committee did not attempt to evaluate the process that selected the specific ERP system or the contractor(s) involved in the development and implementation.

[38]Doug Keene, High Velocity Maintenance, WR-ALC. Personal communication to committee, January 6, 2011.

work package that allows for it. For systems of this age and in these general conditions, a deck of work could be optimized for both cost and turn time. The ALCs over the years have acquired a large and growing workload. This workload reflects the increasing trend of electronics, avionics, computers, and ultimately software in weapon systems. Attendant with an integrated approach, software maintenance became a part of the ALCs and specifically the maintenance organization in the ALCs.[39]

SOFTWARE SUSTAINMENT FOR LEGACY AND FUTURE SYSTEMS

Although not specified in the TOR, the sustainment of software embedded in legacy and future systems was cited as a critical concern by high-level Air Force officials at the outset of the study. The primary focus of the following discussion pertains to the process for software development, through transition and on to sustainment.

Despite the much publicized issues with software development in major weapon systems, relating to cost and schedule overruns, software that is delivered to operational use is generally mature. Software sustainment can be characterized as either (1) the modification or correction of existing code or (2) development of new functions or performance improvements that provide increased capability to enhance weapon system relevance. Problem reports and deficiencies corrections are incorporated into releases that support such capability unless they are urgent enough to ground the aircraft for immediate fixes. Additionally, Diminished Manufacturing Sources (DMS) and support equipment changes may necessitate modifications. Typically, software development and sustainment have nearly identical processes.[40]

Software releases are usually organized in "blocks" where all software changes to the system are developed, integrated, tested, and delivered as a single block release. The Air Force typically maintains a 2-year block release cycle for its aircraft, although this is flexible, and some systems such as ground command and control can have cycles as short as 3 months. Although software is very important for legacy systems, current trends in electronics and software capabilities ensure that future systems will be much more dependent on software to deliver warfighting capabilities. *Growth in software size and complexity in modern weapon systems is and will remain a significant concern for Air Force leadership.* The following topics are discussed in this section: (1) current trends in software development and maintenance; (2) current Air Force capabilities; and (3) future challenges based on new aircraft entering the inventory.

[39] Chapter 2 includes a related discussion under the section entitled *Engineering-based Decisions.*

[40] Karl Rogers, Director, 309th Software Maintenance Group, OO-ALC. "309th Software Maintenance Group." Presentation to the committee, February 1, 2011.

Current Trends in Air Force Software Development and Maintenance

Significant growth in software content across Air Force systems is driven by multiple factors. First, continued increases in the Air Force's dependence on software-intensive systems, and the associated growth in size, complexity, and cost of these systems, have been further compounded by multi-contractor teams using different processes and different tool sets in dispersed engineering, development, and operational locations. Second, software technology is advancing at a breathtaking pace. New technologies and products create both opportunities and challenges that need to be managed during the development and sustainment phases of the life cycle. Third, business and operational needs change, often faster than full system capability can be implemented, which, in turn, challenges the Air Force requirements management processes, existing program management directives, and traditional systems engineering practice. Fourth, software connects other systems in net-centric or system-of-system constructs, increasing the management complexity of software development and modification. Finally, software digitization of previously analog-based, hardware-delivered measurement and control functions, such as sensing, flight controls, and engine controls, continues to increase.

Software presents numerous advantages and challenges to sustainment. First, adopting good software engineering practices often leads to reasonable return on investment; however, investment's benefits may not be realized on initial pathfinder programs. Second, software tends to be invisible in the acquisition process. Typically, the Air Force does not purchase software source code; rather, it buys systems and subsystems that are defined by hardware-centric, prime-item specifications. The total cost of these systems is often significantly underestimated, because the delivered code is but one part, and corresponding support and testing of software can increase costs by significant factors. Third, doing software "right" requires persistence, discipline, and relentless attention to detail. Last, few key decision makers or mid-level managers outside the immediate software community have in-depth software understanding; therefore, software management and engineering competency tend to be weak.

The increased dependence on software is not restricted to DoD systems; it is also evident in commercial products, such as personal digital assistants, smart phones, games, and automobiles. For example, advanced automobiles use upwards of 30 million lines of code, and there exist projections of a 100 million lines-of-code car.[41,42] Indeed, software is the major scheduler and cost driver for the development and maintenance of most systems. Figure 4-12 illustrates many of the above trends,

---

[41]Available at http://www.techweb.com/wire/software/showArticle.jhtml?articleID=51000353. Accessed May 1, 2011.

[42]Available at http://www.eweek.com/c/a/Enterprise-Apps/GM-to-Software-Vendors-Cut-the-Complexity/. Accessed May 1, 2011.

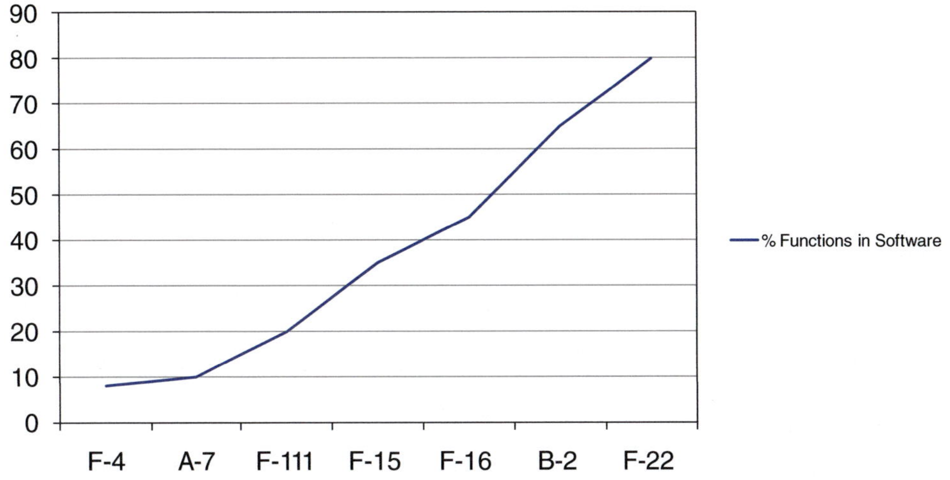

FIGURE 4-12
Increasing percentage of aircraft system capabilities that are now delivered by software. SOURCE: Jack Ferguson. 2001. Crouching Dragon, Hidden Software: Software in DOD Weapon Systems. IEEE Software 18(4):105-107.

showing that an ever-increasing percentage of aircraft system capabilities is now delivered by software. It is estimated that this percentage is quickly approaching 100.[43]

Although code counting and classification methods differ,[44] source lines of code (SLOC) are growing rapidly in modern aircraft, as evidenced in Table 4-3. In addition, the growth in safety-critical SLOC is notable. Complexity of certification for flight-critical software increases rapidly (in a nonlinear manner) as flight-critical code size increases.

As shown in Figure 4-13, the net result is that these factors drive an expected significant increase in dollars consumed to support software.

Air Force Policies for Software Sustainment

Currently, no reasonable, concrete definition of software maintenance exists. There is no set Air Force policy for when software development, upgrade, or modification or when software maintenance begins or ends. During the committee's discussions with software experts and with staff, the common opinion was that

[43]Karl Rogers, Director, 309th Software Maintenance Group, Ogden Air Logistics Center. "309th Software Maintenance Group." Presentation to the committee, February 1, 2011.

[44]Delvyn Deschamps.Air Force Materiel Command. Personal communication to the committee on May 11, 2011.

software is integral, and maintenance, upgrades, modifications, and improvements happen simultaneously. With that as a common approach across the ALCs and with the growth of data previously described, the committee believes that sustainment of weapons system software has the potential to be the largest single growth item in

TABLE 4-3 Software Lines of Code (SLOC) in Selected Legacy and Future Air Force Systems

	F-16[a]	F-22[b]	F-35[c]
Total SLOC[d]	Unknown[e]	5,447,388	19,000,000[f]
Aircraft SLOC[g]	1,710,114[h]	2,730,000[i,j]	9,000,000[k]
Safety-critical SLOC[l]	259	152,000[m]	500,000[n]

[a]Delvyn D. Deschamps, Air Force Materiel Command. Personal communication to the committee, May 11, 2011.

[b]Delvyn D. Deschamps, Air Force Materiel Command. Personal communication to the committee, May 11, 2011.

[c]No data provided by the program office.

[d]Total SLOC—All code associated with the aircraft, including onboard/operational (OFP, fire control, stores management, communications, radar, heads-up display, embedded GPS/ins, digital video recorder, mission planning, etc.), simulators, maintenance and diagnostic software, logistics tracking, etc.

[e]Metric not tracked by the program. Delvyn D. Deschamps, Air Force Materiel Command. Personal communication to the committee, May 11, 2011.

[f]Lockheed Struggles to Keep F-35 Flight-Testing on Track: Report." January 10, 2010. Available at http://www.flightglobal.com/articles/2010/01/21/337424/lockheed-struggles-to-keep-f-35-flight-testing-on-track.html. Accessed May 13, 2011.

[g]Aircraft SLOC—Code for onboard/operational purposes (OFP, fire control, stores management, communications, radar, heads-up display, embedded GPS/ins, digital video recorder, mission planning, etc.).

[h]For current fielded OFP (SCU7/7.1).

[i]Approximate on-aircraft software size (post increment 3.1 with sustainment update 3 merge). SLOC projected to increase to approximately 3.13 million SLOC post increment 3.2. Headquarters U.S. Air Force F-22 Inc 3.2, Modernization Joint Assessment Team Brief, Kathy Watern, SAF/FMC, November 29, 2010.

[j]This number excludes firmware. Margaret Fisher, Air Force Materiel Command. Personal communication to the committee, April 27, 2011.

[k]The article also suggests 11.6 million lines of code as the eventual total. "F-35B Flies with Block 1.0 Software." November 15, 2010. Available at http://www.flightglobal.com/articles/2010/11/15/349732/f-35b-flies-with-block-1.0-software.html. Accessed May 13, 2011.

[l]Safety-critical: A term applied to a condition, event, operation, process, or item whose proper recognition, control, performance or tolerance is essential to safe system operation or use, e.g., safety-critical function, safety-critical path, safety-critical component. Safety-critical computer software Components: Those computer software components and units whose errors can result in a potential hazard, or loss of predictability or control of a system. Available at http://www.system-safety.org/Documents/Software_System_Safety_Handbook.pdf.

[m]Margaret Fisher, Air Force Materiel Command. Personal communication to the committee, April 27, 2011.

[n]Estimated.

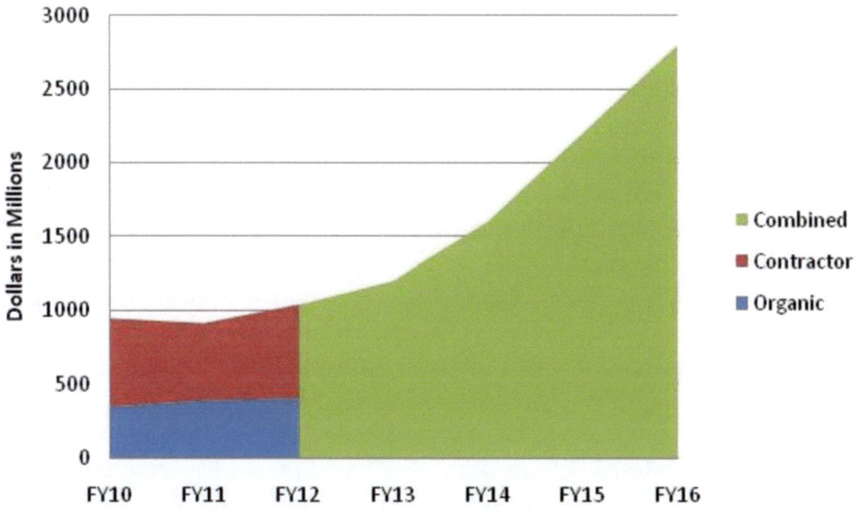

FIGURE 4-13
Software sustainment. SOURCE: Debra K. Tune, Principal Deputy Assistant Secretary of the Air Force for Installations, Environment and Logistics, Office of the Assistant Secretary of the Air Force for Installations, Environment and Logistics. "Developing the Right Product Support Concepts for the Future." Presentation to the committee, October 20, 2010.

the ALC depot maintenance portfolio. As noted in a briefing at OO-ALC, "Even if the number of platforms decreases, software workload will continue to increase."[45]

Software Workforce

Approximately 2,700 people, or 12 percent of the depot maintenance workforce, provide software support to weapons systems at the ALCs. Of these, approximately 90 percent are professionals, and approximately 70 percent (1,800+) have engineering or computer science backgrounds.[46,47] The majority of staff are government civilian personnel, augmented significantly by contractors. The current organizational components are stable, but there is concern over the stability

[45]Karl Rogers, Director, 309th Software Maintenance Group, OO-ALC. "309th Software Maintenance Group." Presentation to the committee, February 1, 2011.

[46]Tom Labrie, 76th Software Maintenance Group (76 SMXG), OO-ALC. "76 SMXG Overview." Presentation to the committee, January 12, 2011.

[47]Karl Rogers, Director, 309th Software Maintenance Group, OO-ALC. "309th Software Maintenance Group." Presentation to the committee, February 1, 2011.

of funding. Funding requirements are generally driven by modernization activities. As individual weapon system programs evolve, the need for software staffing varies.

In addition, as software needs grow, the technical and management workforce must be able to expand with the work. The software workforce can be maintained and grown, but it will be easier at some locations than others. Additionally, the transition from the National Security Personnel System (NSPS) to the standard civil service system will make the task harder.[48] Nevertheless, there is a strong feeling that the needs will be met.

There is strong historical compliance with industry software standards across the ALCs. The WR-ALC and OO-ALC[49] software maintenance organizations are assessed at Carnegie Mellon University Software Engineering Institute CMMI v1.1[50] level 5 (the highest), and both are AS9001/ISO9001 certified.[51] The OC-ALC is assessed at CMMI level 4. All ALCs are currently due to be re-assessed to maintain the currency of their ratings.[52] Each of the ALCs and their software maintenance organizations are critical Air Force assets. They have attained substantial process controls and quality measures at the highest levels of the industry. Their staffs have the technical skill sets required to maintain current inventory aircraft.

Software Facilities

The Air Force has made significant investments in software maintenance facilities at the ALCs. The facilities were comparatively new and largely designed for software maintenance support activities. Although the facilities were generally impressive, the committee noted situations where the equipment and computers used for software maintenance were worn out or suffering from the inability to get repair parts, or the technology had far surpassed the on-hand support equipment. This is particularly important with respect to system or software integration

[48]Under NSPS, the system allowed personnel to be hired at the market rate. In other words, the private-sector pay scale for engineers, for example, could be matched. Under the Federal Employee Retirement System or General Schedule system, a 12, 13, 14 gets what a 12, 13, 14 gets and the added or taking away of the locality pay.

[49]Karl Rogers, Director, 309th Software Maintenance Group, OO-ALC. "309th Software Maintenance Group." Presentation to the committee, February 1, 2011.

[50]Capability Maturity Model Integration (CMMI): A process improvement approach that helps organizations improve their performance. CMMI can be used to guide process improvement across a project, division, or an entire organization.

[51]OO-ALC was assessed CMMI v1.1 level 5 in CMMI V1.3 in 2006. SOURCE: Karl Rogers, Director, 309th Software Maintenance Group, OO-ALC. "309th Software Maintenance Group." Presentation to the committee, February 1, 2011.

[52]OO-ALC is scheduled to be re-assessed for CMMI v1.3 level 5 in October 2011. SOURCE: Karl Rogers, Director, 309th Software Maintenance Group, OO-ALC. "309th Software Maintenance Group." Presentation to the committee, February 1, 2011.

laboratories (SILs). These SILs are major facilities, requiring $100 million or more to stand up and fundamental to system testing.

Finding 4-7. With some exceptions, the ALCs are well resourced to maintain today's existing software. They have achieved quality processes that allow for a strong case for investment to maintain future systems software.

Recommendation 4-5. The Air Force should focus the same, arguably more, attention and investment as that given to equipment in the actual weapon system on tools used for software maintenance. Maintaining currency between test laboratories and actual weapon systems is fundamental for dealing with timing, details of hardware interface behavior, and concurrency.

Organization and Management

A very interesting aspect of the software maintenance organization and processes at the ALCs is the lack of significant organizational and higher Headquarters management oversight. Actually, the committee found this refreshing, even though there are delicate issues requiring higher-level Air Force involvement, albeit not at an excessive level that might stifle local execution. The software leadership of the three ALCs have formed a laudable working arrangement to address workload, process improvement, and policy, fiscal, personnel, and other corporate software issues,[53] but this body has little to no formal authority. Certainly the intent of the AFMC Software Maintenance Group is to provide one virtual software maintenance function, instead of three geographically separated, competing organizations, and a unified message/face to the customer and Headquarters.

Finding 4-8. Weapon system software sustainment is well supported at the three ALCs, but the Air Force has no central governance body to sustain software, and the various Headquarters staffs are technically ill equipped to deal with software issues.

Future Challenges Based on New Aircraft Entering the Inventory

The primary issues with the future software enterprise are driven by two long-standing heuristics: "Software is never finished" and "Software never makes the system cost less." These truisms do not implicate problems with software as a technology, but instead reflect the fact that software can significantly enhance capa-

[53]Tom Labrie, 76th Software Maintenance Group (76 SMXG), OO-ALC. 2011. "76 SMXG Overview." Presentation to the committee, January 12, 2011.

bility and is significantly more flexible to change than physical hardware. Combine that with the growth in dependency discussed earlier in this chapter, and it is easy to conclude that software will be a major and growing cost driver for sustainment of future systems, which must be recognized in planning.

Within the Air Force, better policy and guidance for software development, upgrades and modifications, and maintenance must be developed, promulgated to the field, and enforced. The Air Force would benefit from reviewing its corporate oversight of embedded systems software development and sustainment with an eye toward greater risk identification and mitigation, and from enabling corporate (versus command specific) decision making. In addition, given its dependence on software to achieve mission capabilities, the Air Force would benefit from additional senior leader education in this area.

The Air Force has developed a very strong weapon system software capability, which can be used to improve future weapon system sustainment. Stable workforce planning will continue to be important because the scope of software update manning varies with the amount of modernization in each platform as well as with the scheduling of blocks between platforms.

As cited earlier, the software sustainment processes at the ALCs were considered to be on par with the best of industry. Although process maturity is extremely important, technical expertise in real-time systems, concurrency, new programming languages and operating systems, system of systems, networks, communications, and large-scale integration will be equally important to maintain future aircraft systems. Notwithstanding the process improvement efforts and outstanding performance on various aircraft systems, such as the F-16, the committee is concerned with how the ALCs will build and retain the technical expertise needed in the indicated areas, particularly given the lack of planning for organic maintenance that seems prevalent in the newest weapon systems. Stated another way, there is an overwhelming learning curve to assuming an organic maintenance load when you have not been involved from the beginning of development. Some in the Air Force may prefer long-term contractor maintenance, but at some undetermined point the contractor workforce will become no longer available or extremely expensive. At that point, there will be no fall-back position because the expertise will not be available internally.

> **Recommendation 4-6.** The Air Force should focus on strengthening and retaining the advanced skill sets needed for the sustainment of new aircraft systems.

The facilities and laboratories related to software sustainment are adequate to meet current demands. For future system acquisitions, however, as part of initial program planning, strong consideration must be given to a single government entity or contractor SIL per program. This would localize costs and facilitate transfer of software maintenance responsibilities. In addition, it would generally ensure that the SIL equipment and development tools, which are generally commercial products, are current, as opposed to what sometimes occurs in maintenance organizations; namely, the support equipment is baselined at older versions that are no longer supported by the vendor.

Within Headquarters Air Force, Headquarters AFMC, AFMC Directorate of Logistics and AFMC Directorate of Engineering, there does not appear to be a senior software counterpart that deals effectively with software. In addition, software support for weapon systems is underrepresented (i.e., there is no apparent advocate) in the sustainment community, and software maintenance decisions are defaulted to the program offices and therefore largely to the OEMs. Although this may be acceptable, the Air Force has declared software a core maintenance area under the provisions of 10 U.S.C. § 2464. Because the Air Force believes software to be a core maintenance capability, it must be managed and supported accordingly.

> **Recommendation 4-7.** The Air Force should review its corporate oversight, management, and support of embedded systems software development and sustainment with foci on (1) greater risk identification and mitigation and (2) enabling Air Force corporate, as opposed to command-specific, decision making. In addition, given Air Force dependence on software to achieve mission capabilities, the Air Force should strongly consider additional education on software sustainment for senior leaders.

Access to software development data is vital to proper software sustainment. This includes development of domain expertise with the software and tool sets used in the software development of each platform. Data rights are vital and an issue on virtually all new software workloads; these issues must be considered during acquisition planning to ensure future access for the ALCs.[54] The shift from development to sustainment is currently defined at Milestone C, but it is more a shift from development to sustained block updates—an approach that is not optimum for the new generation of aircraft. A closer partnership between government and contractor throughout the entire development life cycle must be a fundamental part of acquisition planning. Software sustainment planning must begin prior to Milestone A and must have ongoing involvement by the designated sustainment organization throughout the development process. As such, management attention should focus on acquisition/sustainment planning early in the genesis of weapon

[54]Karl Rogers, Director, 309th Software Maintenance Group, OO-ALC. "309th Software Maintenance Group." Presentation to the committee, February 1, 2011.

systems to balance contractor efforts with the required development of organic capabilities.

Recommendation 4-8. The Air Force should pursue a blended partnership between ALCs and contractors throughout the entire software development and sustainment life cycle, commencing before Milestone A.

SUSTAINMENT FUNDING

The Air Force planning and budgeting processes have the capability to successfully address the forecasting and distribution of funds at a high level. The Centralized Asset Management services provided for the Air Force by Headquarters AFMC is a noteworthy improvement. Yet, the budget constraints imposed on the entire Air Force are also felt at the ALCs and impact their abilities to be high-performing, effective, and efficient organizations.

Several limitations on spare parts currently exist because of organizational alignment, material practices, and enterprise funding constraints that have pronounced impacts at the ALC maintenance levels. These involve cost targets, reallocation of priorities for spare parts, and stockage effectiveness targets that fail to realize that parts delays often cause inefficiencies that impact workforce and facility use in a far greater way than the savings created by constraining supplies investment, to say nothing of the direct impacts on material readiness in warfighter operations.

As discussed in Chapter 3, the Air Force has invested in plant and equipment extensively over the past 8 years. Nevertheless, select parts of the plant and equipment do not meet current needs.

Finding 4-9. Despite significant investments, if the ALCs are expected to meet new workloads imported from outside the Air Force repair facilities or to support new technology, additional investment will be required.

As noted early in this chapter, there is a mismatch between funded workload in the maintenance departments and funded support personnel such as engineers to provide technical guidance for the work. This situation can only be understood in the sense that funding accounts do not allow technical engineering support to grow or that policy limits hiring engineering personnel and placing them in the production environment with inherent authority to make decisions. There are disconnects between the engineering resources needed to support production and the resources provided. As discussed in more detail in Chapter 2, the FRCSW has a highly qualified engineering cadre immediately available to support depot production lines.[55]

Although not a funding issue in the traditional sense, the ALCs need an enterprise management solution. The Air Force is slowly fielding an ERP, but a system is needed as soon as practical, and the Air Force must provide it to the entire sustainment community. The impact of not having a system appears to be far greater at the ALCs/AFGLSC because of their total responsibility for fleet support across commands and in some cases across services and foreign military sales.

EFFICIENCY AND EFFECTIVENESS OF AIR FORCE AIR LOGISTICS CENTERS

During the course of the committee's visits, several "white boards," flat screen displays, or paper charts were used to reflect the productivity efforts of the people. Although visual aids are commendable, there was no consistent measure of effectiveness or efficiency across the ALCs or even within an ALC. At the highest levels, there is a lack of measures and objectives for productivity and effectiveness. The committee was not successful in its search for the "key" metric that would tell the chief executive officer of the ALC, AFMC, or the Air Force how the sustainment process or the industrial operation was performing. *A Headquarters AFMC briefing to the committee noted, "Current State Maintenance: AFMC has capacity, personnel and facilities and equipment to sustain legacy systems."[56] The committee disagrees and could not find supporting evidence that legacy systems can be effectively and efficiently supported with current policies and resources. Thus, the committee seriously questions the validity of this statement over the longer term.* The ALCs have strategic plans, and in those the committee observed limited metrics for AA and aircraft on-time delivery from production lines. Although the committee saw displays on aircraft status in the work area and delivery times as a center overview, it could not find evidence that these metrics were being pushed to production teams as a whole. In addition, there were no strong plans for recovery when AA or delivery fell below target levels. The committee observed metrics for quality that exceeded the standard. However, for one or all of these, there was little demonstrated awareness at key supervisor or worker levels. Finally, the committee did not see a metric that related the cost of the production to the planned cost or the planned outcome.

The committee discussed the AA metric at length. The metric has high utility for operational commanders: Do they have the aircraft they need to execute the war

[55]USN. "Fleet Readiness Center Southwest – Roundtable Discussion." Presentation to the committee, March 29, 2011.

[56]Major General Kathleen D. Close, AFMC/A4. "Weapon Systems Sustainment." Presentation to the committee, December 8, 2010.

plans? Therefore, the metric may apply to the overall sustainment community: Is the community at large providing the number of aircraft needed by the war plan? Otherwise, it is not a key performance metric for any manager of the sustainment community. Many argued that the depot maintenance piece, the field maintenance piece, and the supply piece all build the AA metric. There was enough evidence to suggest that the many pieces of the metric result in no one fully taking accountability and responsibility for the performance. The committee universally came to believe that AA is fragmented, and its accountability is such that it is not a measureable performance criterion for any single sustainment manager.

Maintenance depots production metrics were discussed during the visit to the FRCSW where the committee observed excellent quality, cost, and schedule measures. The method of measuring and portraying quality results is much like that in the ALCs. That is, customers are surveyed and the results graphically displayed. Figure 4-14 depicts cost and schedule metrics for two Navy vertical lift platforms.

These metrics immediately resonated with the committee because they portrayed at a glance how the aircraft production lines were doing on cost and schedule. The use of the "0" or planned line as the baseline allows most people to immediately see where the successes have been and what shortfalls have occurred. The charts show performance by each platform as well as by mean performance for the year. Furthermore, these charts are a key part of a bi-weekly Friday standup session with the FRCSW commander that highlights the production results since the last meeting. These standup meetings are held in the quarterdeck of FRCSW and are well attended by many levels of key managers. The charts are constantly on display in a public area.

Finding 4-10. Within the sustainment environment, there is a lack of focus on clear, well-, and widely understood key performance metrics, specifically for cost, schedule, and performance (e.g., cost, delivery schedules, quality), that drive specific actions to improve performance across the sustainment enterprise.

Recommendation 4-9. The Air Force should develop key metrics for sustainment that flow to ALC commanders and that highlight the success or shortcomings of ALC activities, drive appropriate behavior for the workforce, and allow Air Force leadership to assess the health of the enterprise and the adequacy of resourcing for the sustainment process regardless of organizational affiliation.

CONCLUDING THOUGHTS

This chapter evaluated whether the ALCs are adequately resourced in light of the severity of past and current Air Force budgets. As directed by the TOR, the evaluation reaches beyond the mere allocation of monies to support sustainment operations at the ALCs. In fact, the flow of funds for depot maintenance and for the Air Force's flying hour programs has been adequate, although this may not be the case in the near term. Nevertheless, the organizational structure of the ALCs is not resourced adequately in that executive leadership does not have full command or control of the ALC enterprise.

To a serious extent, the supply chain causes great inefficiencies in the depot maintenance and parts repair efforts. Mismatches remain between support to production activities and the growth of requirements at the production level. Importantly, a modern resource management tool, although promised as "coming" is not available, despite being desperately needed. Finally, the only defined measure of effectiveness and efficiency relates to aircraft availability, but far more organizations than ALCs impact aircraft availability. In summary, the ALCs are making it work, but the true full spectrum of resources needed for effective and efficient organizations are not yet available.

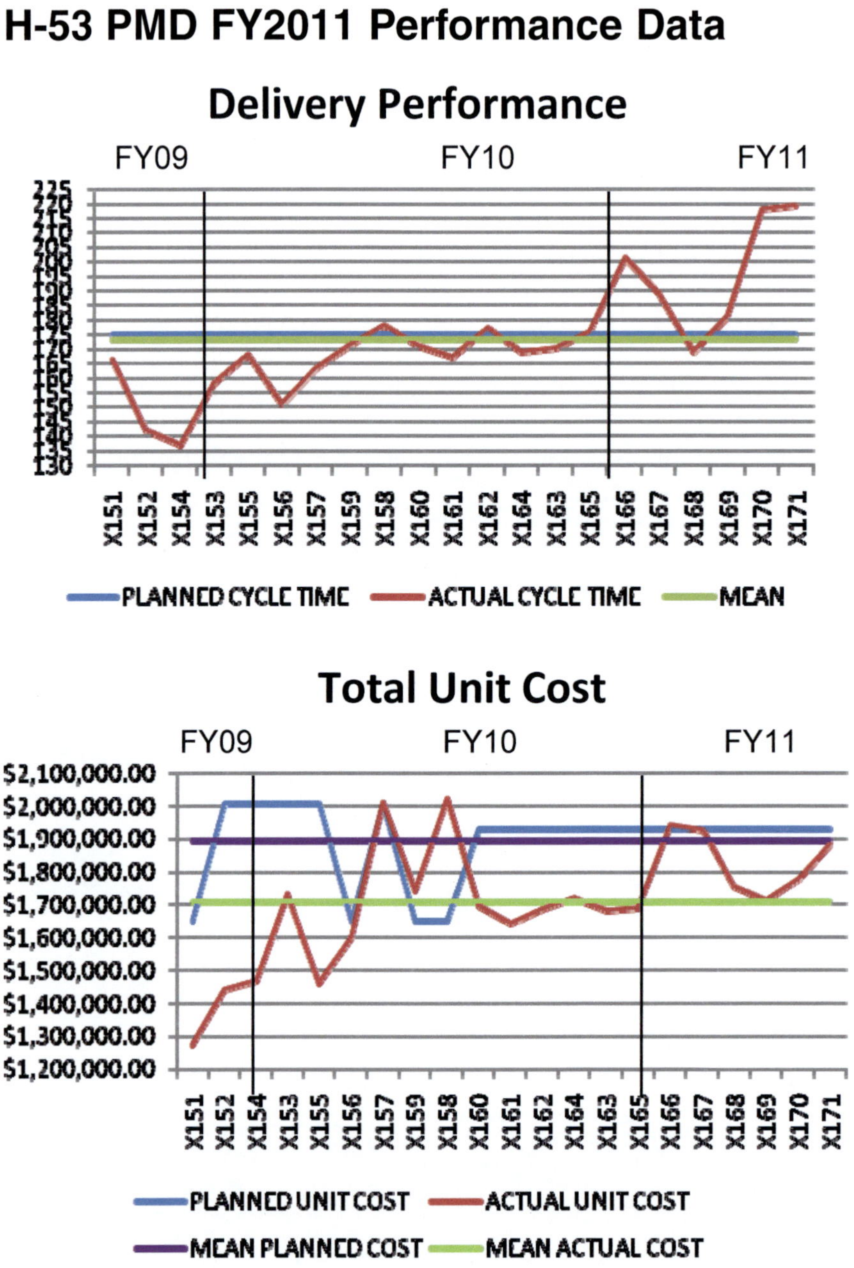

FIGURE 4-14
Cost and schedule metrics for the H-53 and AV-8B. SOURCE: Gregory Mann, United States Navy, Fleet Readiness Center Southwest, Industrial Business Operations Department. Personal communication to the committee on May 16, 2011.

AV-8B PMD FY2011 Performance Data

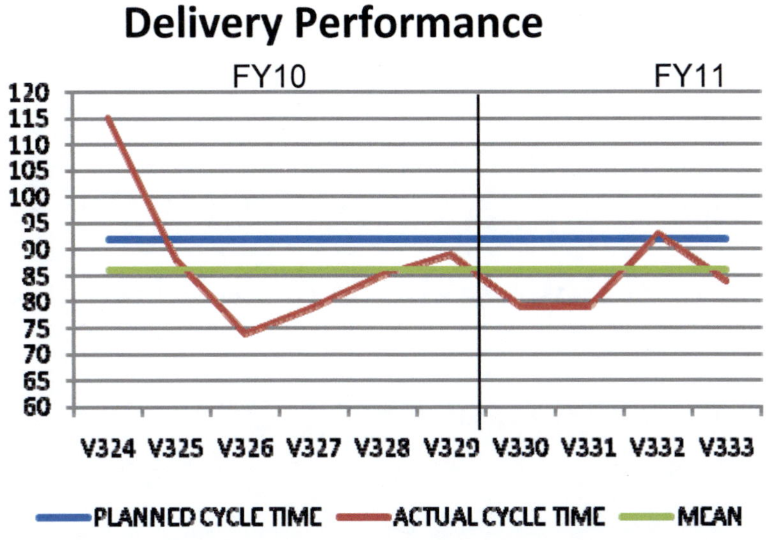

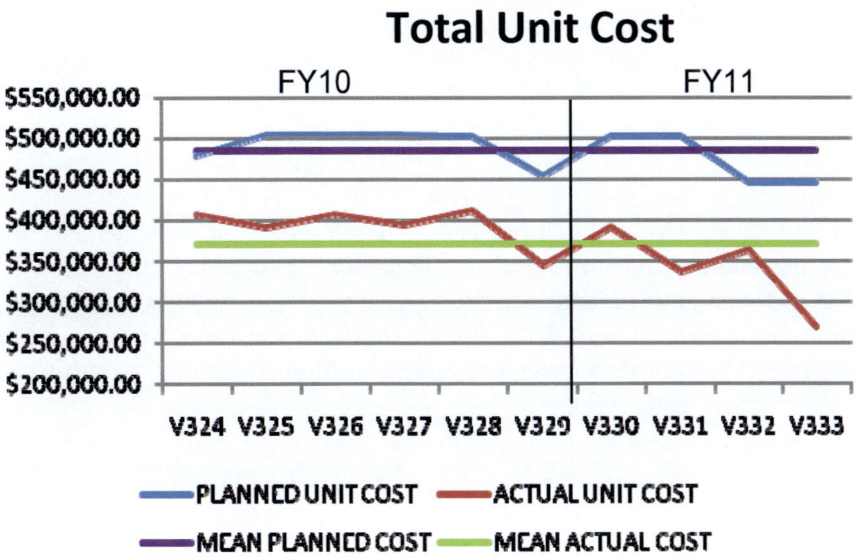

FIGURE 4-14 *continued*

5

Technology Development and Insertion for Sustainment

INTRODUCTION

This chapter addresses element 3 of the terms of reference (TOR), that is, "Determine if any modifications in technology efforts are required, and, if so, identify them and make recommendations regarding the technology efforts that should be pursued, because they could make positive impacts on the sustainment of the current and future systems and equipment of the Air Force." Since its foundation, guided by the prescient words of Gen. Henry H. "Hap" Arnold, the U.S. Air Force (USAF) has demanded that technology development be a key element in providing the wherewithal to make the Air Force second to none in the world. Although the specific organizational structures have varied in response to changing times and needs, research and development (R&D) from the most basic, far-reaching scientific research through development, and on to testing and evaluation have been pursued with vigor.

Academic research, coupled with that of the Air Force laboratories and matured within Department of Defense (DoD)-funded programs and in industry, has led to the extraordinary array of technical marvels that patrol the air, cyberspace, and space domains in times of peace and war. Throughout the Air Force's history, the results of much of this research effort have appeared in new systems, with emphases on immediate advancements in performance. Long-term durability was sought and achieved over time through the application of refined technical understanding and the development and insertion of new materials and processes. New technology also found its way into the maintenance process through the introduction of refined detection apparatus and enhanced repair protocol. In an environment of rapid replacement of old systems with new ones, this strategy has been sustainable with modest maintenance costs. In the present and projected future, however, with limited new system procurement anticipated, a new strategy must govern the introduction of new technology and its impacts on sustainment of the warfighters' requirements to carry out their assigned missions.

The combination of an aging fleet of aircraft with new aircraft whose technology has been primarily utilized to improve the performance of Air Force weapon systems has created a large sustainment cost problem for the Air Force. This problem has been made worse as the size of the Air Forces' fleet has decreased and some aircraft, although small in overall numbers (e.g., the B-2), require a huge sustainment effort to keep them "mission ready." Many examples illustrate how the injection of technology into an existing aircraft system has increased reliability and thereby greatly reduced the sustainment burden of the system (e.g., the F-100 engine required maintenance at 6,000 tacs vs. 4,000 tacs). That said, the non-recurring cost of injecting technology into existing aircraft may impede Air Force acceptance even when the life-cycle cost of not introducing that technology is greater. Although some Air Force technology initiatives have focused on reducing the Air Forces' sustainment burden, in general technology development remains primarily focused on enhanced performance. In addition, programs that historically have been utilized to inject technology into the existing fleet have been weakened or no longer exist.

Although much of the new technology investment in the laboratory is originally targeted at new systems, it may find its way into existing systems. Maintenance depots are increasingly the locale for the insertion of this new technology into legacy systems and those under acquisition but in modification sequence. Sustainment will also continue to be an integral part of new system development as the Air Force focuses attention on its program of Integrated Life-Cycle Management (ILCM).

In this chapter, the implications of ILCM for the technology development and insertion processes will be explored; a broad survey of the array of relevant technologies identified; current technology development and transition processes described and analyzed; and suggestions made for improvement. The breadth of the technical areas and the broad charge outlined in the TOR preclude an in-depth development of a proposed technical agenda; however, recent studies serve to complement this study with respect to comprehensive technical analyses.[1,2,3,4,5]

POLICIES AND GUIDANCE

The task of characterizing the current state of both the technology and institutional and management processes is a daunting one for an institution as complex as the Air Force and its many partners in system development. In addition, the study was conducted in an environment of rapid change in these processes within the DoD and the Air Force in particular. Fortunately, an array of recent related studies has provided background for the committee's particular focus. In the next section, some of these studies will be briefly reviewed to set the stage for the drill-down to the specifics required by the TOR. The following statements regarding life-cycle affordability and sustainment reflect the current vision and responsibilities within senior Air Force leadership:

> Science and Technology **Program Tenets**:
>
> Demonstrate advanced technologies that address affordability by promoting efficiencies, enhancing the effectiveness, readiness, and availability of today's systems, and addressing lifecycle costs of future systems.
>
> S&T **Program Priorities**:
>
> Priority 1.2: Improve the agility, mobility, affordability, and survivability of Air Force assets.
> Priority 2.1: Improve the sustainment, affordability, and availability of legacy systems.

---

[1]Logistics Management Institute (LMI). 2009. Future Capability of DoD Maintenance Depots: Interim Report. LG901M1. December. Mclean, Virginia: LMI. Available at http://armedservices.house.gov/index.cfm/files/serve?File_id=be97f304-3d15-4e96-bc24-689f8cb6c633. Accessed February 20, 2011.

[2]LMI. 2011. Future Capability of DoD Maintenance Depots. LG901M2. February. McLean, Virginia: LMI. Available at http://armedservices.house.gov/index.cfm/files/serve?File_id=394b31e6-4adc-47ca-a6f5-21547f0751fa. Accessed February 20, 2011.

[3]Vince Russo. "Greybeard Assessment of the Sustainment Technology Transition Process." Presentation to the committee, February 7, 2011.

[4]National Research Council (NRC). 2011. Research Opportunities in Corrosion Science and Engineering. Washington, D.C.: The National Academies Press. Available at http://www.nap.edu/catalog.php?record_id=13032.

[5]NRC. 2011. Materials Needs and R&D Strategy for Future Military Aerospace Propulsion Systems. Washington, D.C.: The National Academies Press. vailable at http://www.nap.edu/catalog.php?record_id=13144.

> Priority 4.1: Be a trusted partner of the acquisition/sustainment community to assess technology maturity and enhance and accelerate technology transition.[6]

These priorities must be met within the more specific guidelines that define the acquisition process. Likewise, in AFI-63-101, the Air Force establishes definitions and assigns roles related to technology implications on ILCM.[7] Critical to the current discussion is section 1.4.5, where technology planning and insertion is defined as:

> ...the timely maturation and incorporation of relevant technology throughout the program life cycle to ensure an operationally effective and suitable system. Technology planning and the assessment of technology readiness levels include consideration of such factors as reliability, producibility, testability, *sustainability* and operational performance. Successful technology planning and insertion as part of program life cycle management results in higher fidelity time phased requirements with a more realistic schedule and improved cost estimates [emphasis added]. (p. 12)[8]

Responsibilities for senior leadership in the acquisition chain are clearly articulated. It is the Commander, Air Force Materiel Command (AFMC), who is tasked to "execute the AFMC Mission Assignment Process throughout the ILCM life cycle [and] establish management responsibilities and align the AFMC acquisition and sustainment infrastructure in support of approved missions/levels of service to achieve designated AF ILCM enterprise objectives" (section 2.19.8) and "plan and execute the S&T Program" (section 2.19.14). Interestingly, in defining the responsibilities for the Commander, Air Force Research Laboratory (AFRL), AFI-63-101 does not explicitly mention "sustainment," even though execution of technology development and support of sustainment operations have long been a traditional responsibility of the Air Force science and technology (S&T) laboratories. The Air Force Chief Scientist recently released a 20-year vision document that describes the realm of the possible and dreamed for capabilities that should guide the Air Force's technology development.[9] Embedded in this document are statements that characterize leadership thinking about sustainment:

---

[6]Steven H. Walker, Deputy Assistant Secretary of the Air Force for Science, Technology and Engineering. "Air Force Science and Technology Strategy." Presentation to the Air Force Studies Board, November 16, 2010.

[7]USAF. 2011. Air Force Guidance Memorandum to AFI 63-101, Acquisition and Sustainment Life Cycle Management Incorporating Through Change 3. March 22. Available at http://www.af.mil/shared/media/epubs/AFI63-101.pdf. Accessed April 10, 2011.

[8]Ibid.

[9]USAF. 2010. Technology Horizons: A Vision for Air Force Science and Technology During 2010-2030 (Volume I). May 1. Washington, D.C.: Office of the Chief Scientist of the Air Force.

> Sustainment is essential to the Air Force mission. As legacy air platforms and other systems continue to be used throughout this period, and as new platforms and systems are introduced during this time, technologies to support improved sustainability or to reduce costs associated with sustainment will continue to be essential. (p. 35)

Expanding on this statement of intent, the report clarifies the role of the AFRL Commander by defining a list of actions required of the AFRL:[10]

> 2.1. Determine Alignment of Current S&T Portfolio with "Technology Horizons"
> 2.2. Identify Fraction of Portfolio to be Aligned with "Technology Horizons"
> 3.1. Identify Current Efforts Requiring Realignment or Redirection
> 3.2. Determine New S&T Efforts That Must Be Started (pp. 112-113)

Finding 5-1. The Air Force has recently assigned a higher priority to sustainment technology and has stated its intention to move to an ILCM strategy. Properly implemented, such a strategy implies sensitivity to sustainment in all technology development.

Finding 5-2. Implementation of ILCM is at various stages of development in organizations within the Air Force, but is not yet institutionalized in the research, development, testing, and engineering (RDT&E) system.[11,12]

TECHNOLOGY DEVELOPMENT AND TRANSITION

As suggested in Finding 5-1, to achieve ILCM, sustainment must be built into technology development at all stages. Although the detailed management issues will markedly vary depending on the specifics of the technology and the intended application (maintenance of legacy systems through to development of envisioned systems), broad underlying management issues govern the process of all technology development and transition within the DoD in general and the Air Force in particular. These processes have undergone frequent changes over the past decade or so and are in flux within the Air Force as this report is being written. A recent NRC report described these changing processes in great detail while constraining its focus to new systems. The NRC characterized the program it evaluated from

[10] USAF. 2010. Technology Horizons: A Vision for Air Force Science and Technology During 2010-2030 (Volume I). May 1. Washington, D.C.: Office of the Chief Scientist of the Air Force.

[11] Committee Meeting 2, Wright-Patterson Air Force Base, Dayton, Ohio, December 7-9, 2010.

[12] Committee Meeting 4, Air Force Research Laboratory (AFRL), Tech Edge Innovation and Collaboration Center, Dayton, Ohio, February 7- 8, 2011.

March to August 2010 as one in need of serious attention. Several quotes reflect the NRC's concerns:[13]

> None of the many Air Force presenters to the committee was able to articulate a USAF level, integrated science and technology (S&T) strategy, nor could they identify a single office with authority, resources, and responsibility for all S&T initiatives across the Service. Instead, there appears to be an assortment of technology "sandboxes," in which various players work to maximize their organizational self-interest, as they perceive it. In such a system, optimization will always take place at the subunit level, with less regard for the health of the overarching organization. (p. 7)

> Among the most critical resources are robust processes, from the very conception of a program. For both government and industry, well-defined and well-understood work processes in all phases of program management are essential to successful technological development. Repeatedly during the study, evidence was presented that within the Air Force some of these processes have been diluted in significant ways in the past decade and are only now beginning to be reinvigorated. (p. 29)

> Previous studies suggest that the Air Force needs to do more effective planning in the earliest stages of programs, when ultimate cost, schedule and technical performance are most malleable, and thus most readily influenced. Recently, the Kaminski Report addressed this aspect directly, highlighting the need for systems engineering and the importance of the role that systems engineering plays in the major systems acquisition process.[14] It also persuasively made the case for a return to the days of Development Planning, describing how prior to 1990 the Air Force used Development Planning to assess and integrate the various acquisition stakeholder communities, to include especially combat commands, the Air Force Research Laboratory, and acquisition Product Centers. According to the Kaminski Report, the use of Development Planning, coupled with systems engineering, resulted in the delivery of needed capability to the warfighter in a timely and affordable manner. In addition to Development Planning, there exist two other significant tools in the quest for clear, realistic, trade-off tolerant, stable, and universally understood requirements. These tools are the once-effective ATCs [Applied Technology Councils], in which warfighting commands, acquisition and logistics organizations, and laboratories managed the linkages between operational requirements, technology development, and systems acquisition—with the added benefit of the interpersonal relationships that developed, as well as the face-to-face communications which ensued. The third tool is the establishment and disciplined use of measures of technological readiness, so that only when a technology is well-defined and demonstrated does it make the transition from the laboratory world to become part of a major system acquisition program. (p. 44)

The NRC's 2008 report, supported by extensive benchmarking from successful

---

[13]NRC. 2011. Evaluation of U.S. Air Force Preacquisition Technology Development. Washington, D.C.: The National Academies Press. Available at http://www.nap.edu/catalog.php?record_id=13030.

[14]NRC. 2008. Pre-Milestone A and Early-Phase Systems Engineering. Washington, D.C.: The National Academies Press. Available at http://www.nap.edu/catalog.php?record_id=12065.

technology development and transitions then in place in the United States Navy, United States Army, the Defense Advanced Research Projects Agency, and several industrial concerns, focused attention on the "three R's": requirements, resources, and the right people. It is worth repeating the definitions of these guiding principles because they frame the discussion of the current state of sustainment technology at the close of this chapter:

> 1. Requirements—clear, realistic, stable, trade-off tolerant, and universally understood;
> 2. Resources—adequate and stable, and including robust processes, policies, and budgets; and
> 3. The Right People—skilled, experienced, and in sufficient numbers, with stable leadership. (p. 3)[15]

As dramatic testimony to the fact that the Air Force S&T strategy and implementation are currently in flux, many of the issues raised in the 2011 NRC report had already begun to be addressed by the Air Force at the time of its final publication, and other issues are being addressed by this report. Nonetheless, much remains to be done. In 2010, the Deputy Assistant Secretary of the Air Force (Science, Technology, and Engineering) outlined many of these changes to strategy, development planning, and prioritization by a process similar to the Applied Technology Councils (ATCs) and workforce development. These topics will be reviewed later in this chapter with specific focus on their implications to the development and insertion of sustainment technologies.[16]

> **Finding 5-3.** The Air Force is in the early stages of instituting a focused management approach and of developing plans with requirements, resources, and right people designed to succeed within the ILCM strategy.

TECHNOLOGY AREAS RELEVANT TO SUSTAINMENT

Defining Sustainment Technology Needs

Technology may influence sustainment of the fleet in many ways. These span the range of problem identification and repair of legacy systems to the development of new materials with longer projected lifetimes. Sustainment technology includes those technological advances that, when inserted on an aircraft or aircraft subsystem, produce improvements in the performance life and or maintenance of the

---

[15]NRC. 2011. Evaluation of U.S. Air Force Preacquisition Technology Development. Washington, D.C.: The National Academies Press. Available at http://www.nap.edu/catalog.php?record_id=13030.

[16]Steven H. Walker, Deputy Assistant Secretary of the Air Force for Science, Technology and Engineering. "Air Force Science and Technology Strategy." Presentation to the Air Force Studies Board, November 16, 2010.

aircraft or aircraft sub-system. Sustainment technology may be inserted on aircraft engines, airframe structures, sensors, weapon systems, electronics, hardware, and software. Also included are technologies and advanced processes and practices that include improved forecasting, lean processes and practices, manufacturing, diagnostic and prognostic tools and procedures, personnel education and training, and integrated databases. Too often, technology development programs characterized as "sustainment" are exclusively those targeted at legacy system vehicles, when in fact important sustainment technology opportunities in support of ILCM, including long-term research on underpinning science, may be found throughout the lifecycle. Chapter 4 addresses sustainment of software, while this chapter focuses primarily on vehicles and engines.

In the vehicle and engine areas alone, there is a broad array of relevant technical applications and finite available resources, and a process for their prioritization are required. Recently compiled statistics, shown in Figure 5-1, from the Air Force

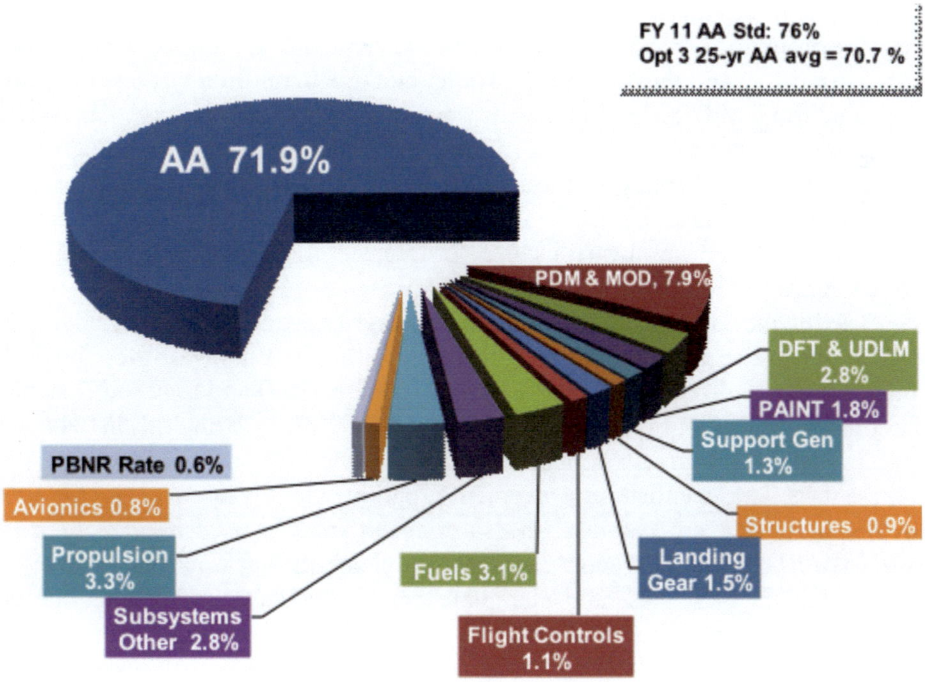

FIGURE 5-1
Total possessed hours for the KC-10 system. AA, aircraft availability; DFT, depot field team; PDM, programmed depot maintenance; MOD, modification; UDLM, unscheduled depot-level maintenance. SOURCE: Fran Crowley, Director, Air Force Fleet Viability Board. "AF FVB Feedback for the Air Force Studies Board." Presentation to the committee, December 7, 2010.

Fleet Viability Board might be considered as input to such a prioritization process. Issues limiting aircraft availability (AA) in the most recent year studied were classified into the several technical root-cause categories listed.

It may be presumed that significant technology fixes might be developed to improve some of the identified issues, but it is clear that no single "fix" will by itself dramatically affect availability of this particular aircraft. Although the S&T community has often been called upon and in many cases has assisted in the development of such point fixes, the preferred strategy is to develop information about issues that are pervasive across platforms and focus development on technologies likely to impact broadly across the fleet. Accomplishments in one such area, high-cycle fatigue, are described in Box 5-1.

In identifying such pervasive technology areas, the S&T community is aided by programs managed within the Aeronautical Systems Center (ASC): for example, the Aircraft Structural Integrity Program (ASIP), the Engine Structural Integrity Program (ENSIP), the Functional Systems Integrity Program (FSIP), and the Engine Component Improvement Program (CIP). CIP also has a history of assisting transitions of sustainment technology, but in recent years has done little of this because of significant budget cuts. With input from the ALCs, appropriate program offices, industry, and the Air Force and other DoD S&T community, these programs continue to identify technology needs, share accomplishments in regular conferences, publish standard practices, and in the best of circumstances influence the identification of priorities for funding by one or more entities. Figure 5-2 depicts one of the key areas addressed by ASIP; the committee was informed that laboratory work is under way at various levels of intensity in each of the areas indicated in the figure.[17]

The interaction of the AFRL with these integrity programs is explored further below, but in the present context, it is worth noting that many of these integrity programs have a long history (ASIP was initiated in 1958) and represent a significant resource for identifying "needs" to be addressed. Translating this list of needs into funded "requirements" is a complex process involving many other players. These several independent improvement programs identify opportunities in their own spheres, *but no comprehensive sustainment technology plan currently exists within AFRL.* In its 1997 report exploring the elements that would be expected in such a plan, the NRC presented a template that might be used as AFRL reexamines its sustainment portfolio:

1. Develop an overall strategy that addresses the Air Force aging aircraft needs
2. Recommend and prioritize specific technology opportunities in the areas of

[17]Pam Kobryn. "Aircraft Structural Integrity Program (ASIP)." Presentation to the committee, February 7, 2011.

BOX 5-1
High-Cycle Fatigue (HCF)

The solutions to the HCF problem that plagued the Air Force for nearly a decade could not have been possible without the contributions of the Air Force S&T community.

The combination of high frequencies (up to 1,000 Hz), millions of cycles, and low amplitudes can result in HCF.[1] The presence of manufacturing defects or Foreign Object Damage (FOD) provide sites from which cracks can grow as a result of many millions (or billions) of cycles at stresses well below the yield strength of the material.[2] FOD provides the source of initiation of the crack. HCF provides the method of propagation. If not detected in time, the end result is catastrophic failure of the component. Combining HCF conditions with the increasingly higher performance provided by advanced materials and designs exacerbated the phenomenon.[1]

From 1995 to 2003, HCF was the major contributor to the failure of components in military gas turbine engines.[3] Studies of the rates of Air Force mishaps over a period of 15 years showed that more than 50 percent resulted from HCF. Similar data for the United States Navy showed that more than 40 percent of mishaps resulted from HCF. Also during this time HCF began to appear in commercial engines to a lesser extent than in military engines but with severe consequences to the manufacturer's development programs and revenue service for airlines.[1] HCF affected virtually all engine components and many of the materials. It impacted not only engine reliability and safety of flight, but also sustainment, requiring increased field inspections and depot maintenance and reduced aircraft availability.

This problem became so pervasive that a major program was initiated to solve it. The AFRL began the HCF Initiative with the strategy of developing the tools and techniques to change the basis of HCF design from empirically based to physics based and then to demonstrate and transition these tools to the industry design systems. The Initiative consisted of seven action teams: Materials Damage Tolerance, Component Surface Treatment, Passive Damping Technology, Forced Response Prediction, Component Analysis, Aeromechanical Characterization, and Instrumentation. Engine demonstrations were an eighth action team, but as the engine demonstrators supported the overall Integrated High Performance Turbine Engine Technology program, they were eventually not counted in the HCF Initiative.[1]

The impact of the HCF Initiative on current and development engines was enormous. The field engine inspection workload for HCF was reduced by more than 90 percent, and the proportion of engine mishaps resulting from HCF was reduced from 54 to 7 percent, far exceeding the HCF program goal of 50 percent.[1] Further, these same tools became enabling technologies for the next generation of high-performance jet engines, including the F-135 engines for the F-35 Joint Strike Fighter. Without these tools and methods, the development programs would likely have encountered many unexpected difficulties with the accompanying delays and cost growth.[1]

[1]Theodore Nicholas. 2006. High Cycle Fatigue, A Mechanics of Materials Approach. London, UK: Elsevier.
[2]Danny Eylon, University of Dayton, personal communication.
[3]B.A. Cowles. 1996. High cycle fatigue in aircraft gas turbines—an industry perspective. International Journal of Fracture 80:147-163.

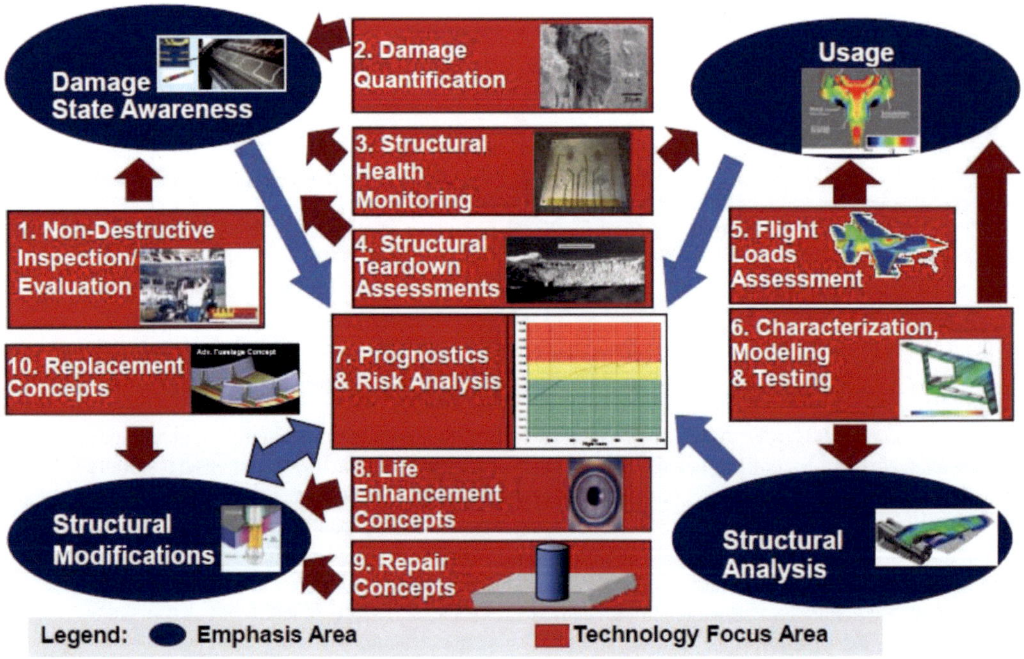

FIGURE 5-2
Condition-based maintenance as part of the Aircraft Structural Integrity Program. SOURCE: C.A. Babish IV, "ASC/EN." Presentation at Panel Session Kick-Off, 2008 ASIP Conference.

- fatigue, corrosion fatigue, and stress corrosion cracking
- corrosion prevention and mitigation
- nondestructive inspection
- maintenance and repair
- failure analysis and life prediction methodologies[18]

The 1997 NRC report described in great detail the state of the art of all of the above and suggested 49 specific technical recommendations. This report was well received by the Air Force and was influential at the time in determining not only the research agenda, but also the strategies for inserting technology into practice. This report offers an appropriate template for the sustainment tasks ahead for

[18]NRC. 1997. Aging of U.S. Air Force Aircraft: Final Report. Washington, D.C.: The National Academies Press. Available at http://www.nap.edu/catalog.php?record_id=5917. Accessed November 22, 2010.

AFRL, described more recently in *Technology Horizons*[19]; namely, a comprehensive examination of needs and opportunities that will guide investment strategy within the S&T arena. This process of developing an investment strategy is currently under way, assisted by a panel assembled by the Air Force Scientific Advisory Board (AFSAB) and charged with tasks including some that are in many ways similar to those performed by the Ageing Aircraft Study in the 1990s. The results of the AFSAB study will be briefed to the Air Force in June 2011.

Long-Term Research

In any discussion of technology, it is important to note the role of long-term research. Long-term ("basic" or "fundamental") research in the technical areas identified in the current report have been funded over the years by many agencies including the National Science Foundation, Department of Energy, National Institutes of Standards and Technology, Federal Aviation Administration, and others, but a great deal of the specific focus on Air Force-related problems has come from efforts of the Air Force Office of Scientific Research (AFOSR), the Office of Naval Research, Army Research Office, and the National Aeronautics and Space Administration. The following sections summarize the results of several key studies that explore the status of these support S&T areas.

Corrosion

Corrosion, broadly defined as atmospheric degradation of materials, is ubiquitous. Although rarely causing catastrophic failure to systems because it is found and repaired, corrosion adds to sustainment costs and extended depot time. Any improvements in corrosion resistance in new materials, mitigation and/or detection in old materials, repair technology, and life prediction promise to be money saving and enabling of fleet readiness. The subject of corrosion and its interaction with mechanical degradation received significant attention in the 1997 NRC report discussed above, and was attacked with some vigor by the Air Force in the years immediately before and after that report's release. In recent years, corrosion is no longer the center of Air Force or other federal agency attention, even though significant new tools and opportunities for understanding corrosion have matured in related fields of science and engineering. In 2010, the NRC released a broadly based study on Research Opportunities in Corrosion Science and Engineering

[19]USAF. 2010. Technology Horizons: A Vision for Air Force Science and Technology During 2010-2030. Volume I. AF/ST-TR-10-01-PR. May 15. Available at http://www.af.mil/shared/media/document/AFD-101130-062.pdf. Accessed May 25, 2011.

(ROCSE), outlining the needs and opportunities in this important field. Quoting from this report:

> There are many forms of corrosion, and while some are well understood at the macro level, complex interactions among the different forms are yet to be fully clarified. Further, at the detail level there is often relatively poor understanding of corrosion mechanisms, which makes it technically difficult to devise cost-effective engineering solutions to predict, avoid, and mitigate corrosion damage. These difficult problems have often been put on hold in favor of short-term, empirical fixes, but now appears to be an opportune time to readdress complex questions with new techniques. Advances in characterization (using, among other techniques, transmission1 and scanning electron microscopy, micro- and nanometer electrochemical probe methods, synchrotron beam lines and lasers, x-ray, and neutron spectroscopy and combinations of these methods for simultaneous information gathering) and computation and modeling (first principle, molecular dynamics, multiscale modeling, and informatics) have dramatically broadened the array of tools available.

ROCSE recommends that all federal agencies develop and carry out a targeted plan for corrosion research. It is anticipated that near-term impacts on corrosion mitigation and management may be achieved, and that long-term effects will be seen in the development of new materials with less need for such costly repair.

Propulsion Materials

The relative ease of re-engining offers unique opportunities for technology insertion into aging systems. Although the principal drivers for such engine replacements are reductions in fuel costs and/or improvement in performance, reduced sustainment costs may also be an immediate consequence of re-engining. A 2006 NRC report broadly looked at all DoD aerospace propulsion needs and focused on how technology may improve performance, specifically thrust-to-weight and fuel consumption.[20] Although it gave very little direct attention to sustainment, the NRC noted that the further development and insertion of new materials such as superalloys, composites, and ceramic matrix composites that have longer life, increased fatigue resistance, and improved temperature resistance would impact sustainment. Specific recommendations relevant to the current study include:

> 3-1. To accelerate the development of new engine technologies, the Air Force gas turbine S&T funding should be increased significantly, from approximately $100 million annually to a level that reflects buying power at the time when the F-15 and F-16 engines were being developed. Top priority should be given to overcoming the technology barriers that will have the largest impact on future weapons systems:

[20]NRC. 2006. A Review of United States Air Force and Department of Defense Aerospace Propulsion Needs. Washington, D.C.: The National Academies Press. Available at http://www.nap.edu/catalog.php?record_id=11780.

- Compressor discharge temperature limits,
- Turbine inlet temperature limits,
- High-temperature, high-heat-sink fuels for thermal management,
- Lightweight structures, and
- Signature control. . . .

3-3. DoD should restore gas turbine S&T funding under the Versatile, Affordable, Advanced Turbine Engine (VAATE) program to the original planned level. VAATE should address the primary risk areas necessary to advance jet engine technology, which include a robust engine demonstrator program and key producibility challenges. . . .

3-5. DoD should reinstate an engine model derivative program (EMDP) to speed the transitioning of technology to the legacy fleet to improve safety, reliability, and *affordable readiness* for DoD. An earlier EMDP demonstrated its utility and value for the current fleet of engines, most of which were developed spirally through this program or similar programs in the commercial sector [emphasis added]. . . .

3-9. DoD should invest in several critical technologies that will impact all types and classes of propulsion systems: high-temperature materials, including high-temperature blade/vane materials and coatings; high-temperature and high-heat sink fuels; lightweight structures; and accurate analytical modeling.

Integrated Computational Materials Engineering

Introduction of new materials into the design space presents one of the greatest opportunities to enhance sustainment through longer projected system life. Additional drivers for materials substitution include enhanced performance and, increasingly, issues of availability of scarce alloying elements (rare earths and others) and environmental concerns (e.g., elimination of corrosion-inhibiting chromates from paints). In recent years, engineering practice, aided by an array of computational tools, has shortened the design to manufacture time while new materials development has lagged behind. Constrained by a largely empirical strategy for development and testing, new materials are frequently "not ready" for inclusion in the designer's portfolio and are always considered the "material of the future."

Recent developments, building on years of basic research in fundamental understanding, computation, and experimental technology, have led to a new paradigm for materials development, Integrated Computational Materials Engineering (ICME). When properly exploited, ICME shortens the time and dramatically reduces the cost of new materials development. A recently released NRC report (the "ICME report") reviews the history of these developments, describes the current state of the art, and lays out an agenda for further R&D. Properly implemented, ICME promises rapid development of new materials with the same or better properties, at the same or lower cost as substitutes with availability threatened by environmental concerns or specialty alloy requirements, and in adequate time to address these pressing concerns as legacy or new systems demand modification. Expanded efforts in ICME-enabled materials development are now in place in all three service laboratories, targeted at issues specific to each service's needs, and supported by state-of-the-art understanding of science, including that required to extend useful life in service. In summary, science and engineering developments offer opportunities to reduce cost and extend useful life for both legacy and future systems.

Finding 5-4. Targeted technology development and insertion have long been critical elements in addressing sustainment needs within the Air Force. Advances in the underlying sciences continue to offer new opportunities for development of technologies that would support the Air Force goal of ILCM. Several independent improvement programs identify opportunities in their own spheres, *but no comprehensive sustainment technology plan currently exists within AFRL.*

Addressing the development and transition of technology requires a robust system of prioritization, establishment of requirements, proper resourcing, and an adequate workforce to execute planned programs. The following section discusses the specifics of these elements within the Air Force technology development and transition process, beginning with a brief review of recent years.

AIR FORCE SUSTAINMENT TECHNOLOGY

Overview

Sustainment technology that is inserted on aircraft and aircraft sub-systems involves several developmental centers/organizations including AFRL Technical Directorates (TDs) and industry original equipments manufacturers (OEMs). Because an OEM possesses first-hand manufacturing knowledge, its relationship with the aircraft or aircraft sub-systems is strong, creating an advantage in developing sustainment technology that is relatively easier to insert into the aircraft and aircraft sub-systems. OEM-driven block upgrades and product improvements of aircraft and aircraft sub-systems include programmed developments of sustainment technology. These types of sustainment technology development may continue over an extended portion of the aircraft life cycle.

The control of insertion of sustainment technology into aircraft and aircraft sub-systems may involve several stakeholders including industry OEMs, Air Combat Command (ACC), Air Mobility Command, AFMC, ASC, and the Air Logistics

Centers (ALCs). Once the technology is developed to a Technology Readiness Level (TRL) of 6 or higher within an AFRL TD, the System Program Office (SPO), OEM, and/or the ALC may control the insertion of sustainment technology. The age of the aircraft or aircraft sub-system is an important variable in determining which organization has authority to order sustainment technology insertion. In this complex system with many players), trust and respect between the supply side (AFRL TDs, OEMs, SPOs) and support side (ALCs) are critical for effective and efficient decision making to insert sustainment technology. However, success depends not only on trust and respect but also on well-defined strategies and processes.

Within the Air Force, the AFRL TDs are the primary source for the development of sustainment technology. The following sections discuss the recent history of sustainment development within the AFRL, in addition to the often-changing strategy and processes in which it has been lodged during the past decade.

Historical Background

The Air Force S&T program has undergone many organizational changes since the creation of the Air Force. The current organization, the Air Force Research Laboratory, evolved from several individual labs that had strong historical linkages with different centers and commands, driven largely by their specific technology focus and by geographic proximity. Many of these linkages continue to this day and constitute an important element in identifying the technology agenda and transition pathways. Twenty-five years ago, the Air Force S&T program was housed in 13 labs plus the Rome Air Development Center in New York.[21] In 1990, these 13 labs, plus Rome, were reorganized into four major laboratories—Armstrong Laboratory, Phillips Laboratory, Rome Laboratory, and Wright Laboratory.[22] These four "superlabs" were aligned with four Air Force product centers—Wright Laboratory with the Aeronautical Systems Center, Phillips Laboratory with the Space and Missiles Systems Center, Armstrong Laboratory with the Human Systems Center, and Rome Laboratory with the Electronics Systems Center. Each superlab had missions and investments supporting these centers as well as other parts of the Air Force.

The formation of AFRL in 1997 consolidated all of the Air Force S&T efforts into a single organization reporting to the AFMC commander. The new organization, consisting of key headquarters (HQ) functions and 10 technology directorates, one of which is the AFOSR, is shown in Figure 5-3.

AFRL's website describes its mission as "leading the discovery, development and integration of affordable warfighting technologies for America's aerospace

[21] Robert W. Duffner. 2000. Science and Technology, The Making of the Air Force Research Laboratory. Maxwell Air Force Base, Alabama: Air University Press. November. p. 12.

[22] Ibid.

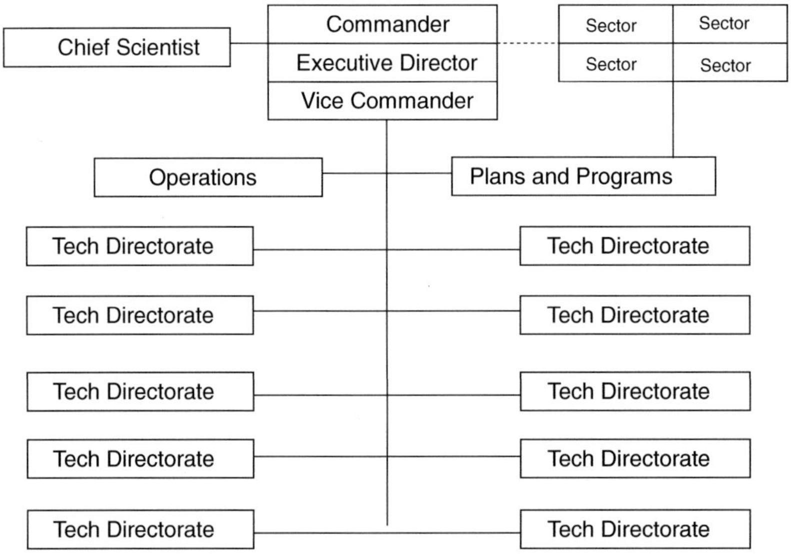

FIGURE 5-3
Early notional organizational structure of the Air Force Research Laboratory. SOURCE: Robert W. Duffner. 2000. Science and Technology, The Making of the Air Force Research Laboratory. Maxwell Air Force Base, Montgomery, Alabama: Air University Press.

forces. It is a full-spectrum laboratory, responsible for planning and executing the Air Force's science and technology program. AFRL leads a worldwide government, industry and academia partnership in the discovery, development, and delivery of a wide range of revolutionary technologies. The laboratory provides leading-edge warfighting capabilities keeping our air, space and cyberspace forces the world's best."[23] To support this mission, AFRL develops an investment strategy, delineating its emphasis areas, including near-, mid-, and long-range technologies. The eight TDs are responsible for discovering, developing, and transitioning the technologies required by this strategy. Out of necessity, the strategy changes with time. World events, emergence of revolutionary technologies, and Air Force strategic changes can impact the S&T strategy. The TDs' technology programs change to keep pace with the current strategy.

Customer interface was a strong headquarters focus during AFRL's initial startup. As shown in Figure 5-3, an early construct of AFRL had sectors—representing major customer focus/investment areas—led by Colonels, interfacing directly with the Commander. Six sector offices—Aeronautics, Space and Missiles, Com-

[23] Available at http://www.afrl.mil. Accessed May 4, 2011.

mand and Control, Human Systems and Logistics, Weapon Systems, and Modeling and Simulation—eventually were formed within AFRL. The sectors' focus was three-fold: customer interface via a single point of entry, integration across TDs, and investment strategy. To assist with customer interface, AFRL representatives served as collocates in the major commands (MAJCOMs). These representatives reported to their various sector chiefs. Even though the intent was that the sectors would be a single point of entry, customer technology transition processes were numerous, diverse, and spread across each of the TDs.

Many transition processes involved relationships that existed prior to the AFRL's formation, especially with the product centers. Some were the direct result of the TD's geographical proximity to a major Air Force unit, such as the proximity of several AFRL TDs to ASC. Many resulted from the TD's direct interface with its customers, in some cases supported with major 6.3 programs. Many were the result of strong ties to industry, such as that of the Propulsion Directorate to engine companies. Many "transition programs" lacked formal agreements or documents that identified key performance parameters or that put critical resources in place to ensure the transition. The formation of the sectors was one way to improve the transition process and take some of the customer interface/integration workload off of a hard-pressed staff.

A good example of informal relationships with a sustainment focus may be found in the programs of the Systems Support Division in the Materials and Manufacturing Directorate. This division maintained the expertise and facilities to do quick reaction support to urgent Air Force sustainment issues. This division also had collocated engineers in each of the major SPOs, such as F-22 and C-17, to identify issues and bring S&T solutions to the needed sites. Customers would frequently go directly to the quick reaction team or use the collocated engineer as a point of entry to directorate personnel and technologies. The division also housed program offices—corrosion, non-destructive evaluation, and composites—at each of the ALCs. Other TDs also had their own informal structures, often involving collocated personnel at major stakeholders' offices. In essence, a network was in place to provide a multitude of sustainment solutions.

When AFRL was established, its technology portfolio was as diverse as its TDs, spanning basic research, new materials, advanced propulsion, sensors, and human resources. Within this diverse portfolio, however, sustainment was a significant part of the investment strategy and transition focus. The Aeronautics Sector had sustainment as a major part of its portfolio. Four TDs—Materials and Manufacturing, Propulsion, Air Vehicles, and Human Effectiveness—had major sustainment investment areas. Each TD had numerous examples of successful transitions that impacted sustainment cost/aircraft availability. The sectors were key participants in many of these successful transitions. Two TDs, Materials and Manufacturing and Air Vehicles, jointly manned/resourced an aging aircraft office whose mission was to call increased attention to aging aircraft issues with technology solutions. This office was closely aligned with an aging aircraft office in ASC, which was formed shortly after the release of the 1997 NRC report.[24] A special 3600 Program Element (PE) was created specifically for the aging aircraft function. Eventually this PE was transferred to the ASC aging aircraft office, where it became a key source of funds to facilitate the technology transition of sustainment-related technologies to the ALCs. The office was also strongly supported by the ASIP personnel in ASC. In those early years, the aging aircraft offices, the Aeronautics Sector, and the ALCs had substantial collaborations that produced significant advances for Air Force weapon system sustainment. Many technologies transitioned through this collaboration, including those related to paints/coatings, wiring, corrosion control, composite repair, and composite patches.

The FY2000 S&T plan for AFRL shows sustainment as a key emphasis area, supported by the Aeronautics Sector, the relevant TDs, and an aging aircraft office. The 1997 NRC report was a key reference for the 2000 plan.[25,26] Even though the plan is more than a decade old, sustainment was already recognized as a major developing need, forecasted to increase dramatically in importance over the coming years to maintain air worthiness for aircraft being used past their anticipated service life. The plan also recognized that sustainment was not just an aging issue. Advanced systems entering the inventory frequently created sustainment issues, such as low observable (LO) maintainability, which required new inspection and repair concepts. However, the level of emphasis that sustainment would receive was to wax and wane in the next decade as S&T strategy and processes changed.

Changing Strategies and Processes

2000-2002

Shortly after AFRL's creation, the Air Force S&T investment strategy experienced a change in emphasis that had a major effect on sustainment. The 2000 strategy called for, among other things, a migration of focus from aero to space, development of enabling technologies, and identification of an Integrated Technology Solutions (ITS) strategy to address high-priority warfighter needs.[27] The goal of the ITS strategy was to mesh the technical efforts of multiple TDs into Integrated

[24]NRC. 1997. Aging of U. S. Air Force Aircraft: Final Report. Washington, D.C.: The National Academy Press. Available at http://www.nap.edu/catalog.php?record_id=5917.

[25]Ibid.

[26]USAF. The Air Force Science and Technology Plan Fiscal Year 2000. Available at http://www.wslfweb.org/docs/st_plan00final.pdf. Accessed May 5, 2011.

[27]USAF. The Air Force Science and Technology Plan Fiscal Year 2000. Available at http://www.wslfweb.org/docs/st_plan00final.pdf. Accessed May 5, 2011.

Technology Thrusts (ITTs), providing a synergistic laboratory solution. Each ITT contained several Integrated Technology Thrust Programs (ITTPs) supported by programs in the TDs. One of the initial six ITTs was aircraft sustainment. Within this ITT were four ITTPs directed at aging aircraft structures, LO maintainability, HCF, and turbine engine durability.[28] Each ITTP supported a documented critical warfighter operational need as defined in the Air Force Modernization Planning Process (AFMPP) and needed to align with Air Force core competencies.[29]

The enabling technologies strategy was implemented through programs in the individual TDs. Several sustainment-related enabling technology programs were being implemented in the TDs during this time period. For example, materials and manufacturing had LO maintainability, paint, aging aircraft, HCF, pollution prevention, and nondestructive evaluation (NDE); air vehicles had aging aircraft; and propulsion had HCF, to name a few. Transition of the technology being developed under the Enabling Technology efforts was accomplished through arrangements set up by the TDs or by the TDs combined with the sectors.

Clearly, sustainment was a major emphasis area for the AFRL during this period, and successes were numerous. A particularly noteworthy success was the solution of the HCF fatigue problem (see Box 5-1), which affected virtually every fighter aircraft engine in the Air Force inventory. The combination of outstanding technical expertise and leadership, TD programs, ITTs, and partnerships with industry led to the successful solution of this major showstopper.

2002-2004

During 2002-2004, Long-Term Challenges and Short-Term Objectives were identified to support Air Force Vision 2020.[30] A Short-Term Objective had to address compelling Air Force requirements, have strong user support, and attain its objectives within 5 years at a TRL of 6. There were eight Short-Term Objectives, with one being sustaining aging systems.[31] One of the areas recommended for increased emphasis at the applied research level was Integrated Vehicle Health Management (IVHM) directly impacting sustainment. IVHM uses a systems approach and is designed to detect and diagnose the condition of an air vehicle to determine the usable safe operating life and/or the need for any maintenance actions. Within this system, using a strictly Condition-Based Maintenance (CBM) approach versus a Programmed Depot Maintenance (PDM) approach would be a major capability

---

[28] Ibid.

[29] Ibid.

[30] Don Daniel. "AF S&T Investment Strategy and Funding." Briefing to NDIA, February 12, 2002. Available at http://www.dtic.mil/ndia/2002science/daniel.pdf. Accessed May 6, 2011.

[31] Ibid.

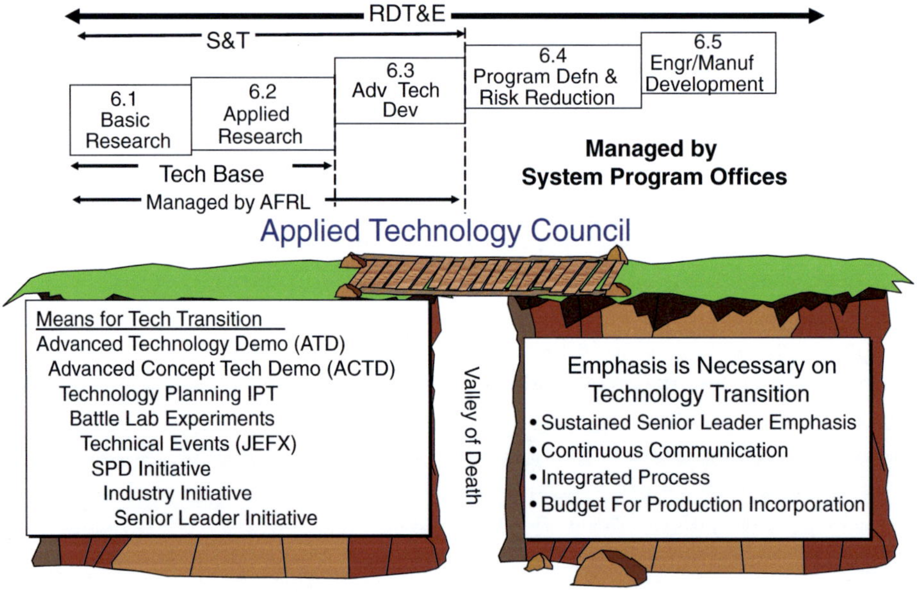

FIGURE 5-4
The Applied Technology Council (ATC) was designed to link the S&T community with the user community. SOURCE: Major General Paul D. Nielsen, Commander, Air Force Research Laboratory. "AFRL Overview Briefing." May 2003.

afforded the sustainment community with positive aircraft availability results. Key technologies for development were sensing methodologies, signal processing, and prognostics. New initiatives were identified that had major implications on next-generation warfighting capabilities, such as information technology, biotechnology, nanotechnology, space technology, and directed energy. New, revolutionary technologies aimed at sustainment were not in this mix.[32]

The major process change during this time period was the creation of the ATCs. The ATCs were organized to address the many difficulties that were severely hampering the transition of technologies from the AFRL to its customers. In many cases, funding shortfalls, funding reallocations, requirement changes, shortfalls in key performance parameters, and personnel changes occurred without the knowledge of key stakeholders. These pitfalls were viewed as the "valley of death" between the S&T community and the user community and are schematically represented in Figure 5-4.

As indicated in Figure 5-4, there were many paths for technology transition, but

[32]Major General Paul D. Nielsen, Commander, Air Force Research Laboratory. "AFRL Overview Briefing." May 2003.

they required the emphasis areas shown on the right to successfully navigate the valley. To achieve this emphasis, ATCs were established as semi-annual meetings of the AFRL Commander, the Product Center Commander, and the Vice Commander of the receiving MAJCOM. It was through this process that technology "needs" as seen from the AFRL perspective would become technology "requirements" as seen from the warfighter perspective. The primary outputs of the ATCs were programs or groups of programs commissioned as Advanced Technology Demonstrations (ATDs) by the senior leadership of the three entities noted. ATDs were binned in four categories:

1. Category 1: MAJCOM/Agency supports and has programmed required funding for transition (6.4 and beyond) within the Future Years Defense Program (FYDP)
2. Category 2A: MAJCOM/Agency supports and is committed to identify transition funding in the next programming cycle
3. Category 2B: MAJCOM/Agency supports but is not currently able to Program Objective Memorandum(POM) for transition
4. Category 3: Warfighter does not support[33]

Each round of an ATC was focused on a specific MAJCOM, with ATC #6 focused on AFMC. Four Category 1 ATDs that supported sustainment were commissioned from ATC #6:[34]

1. Bonded Repair Capability Enhancements
2. Corrosion Effects on Structural Integrity
3. Advanced Aircraft Corrosion Protection
4. Advanced Non-destructive Evaluation (NDE) for Aging Systems

A further ATD on LO maintainability was established in the ATC with ACC. In addition to the ATDs, the TDs continued to support sustainment through their individual programs, quick reaction capabilities, and collocated engineers. However, their level of support was dramatically reduced in the aging aircraft area. Largely because of budgetary pressures, the AFRL and ASC eventually closed their aging aircraft offices.

[33]Major General Paul D. Nielsen, Commander, Air Force Research Laboratory. "AFRL Overview Briefing." May 2003.

[34]Mike McMillan. "Air Force Technology Transition and Modernization." Briefing to JTEG, 2001.

2005-2009

The next major change in strategy and process centered around the creation of Focused Long-Term Challenges (FLTCs). During the 2005-2009 timeframe, the S&T program was strongly influenced by the demands of the 21st century environment. The Global War on Terrorism (GWOT), cyber warfare, and precision strike, to name a few, were major drivers not envisioned in earlier years. The S&T program was also under continuing pressure to be more innovative and forward thinking, especially with the GWOT changing strategies and tactics. These mounting pressures resulted in the AFRL creating an "integrated capability-based planning and programming" approach. Also during this timeframe the Air Force adopted a new technology vision that AFRL had already embraced: Anticipate, Find, Fix, Track, Target, Engage, Assess—Anything, Anywhere, Anytime (AF2T2EA4).[35] This technology vision provided a link to the Air Force's capabilities-based planning and Capability Review and Risk Assessment (CRRA) process. The S&T program was binned in three time-phased groupings:[36]

- Rapid Reactions—technology options for immediate warfighter needs
- ATDs—technology options that meet near-term needs
- FLTCs—technology options for innovative capabilities for future needs

The planning and implementation of the FLTC process and portfolio became a primary emphasis of AFRL during this period. Eventually, this concentrated effort produced eight FLTCs. One of these—FLTC #8: Affordable Mission Generation & Sustainment—was the home of sustainment technologies for the future Air Force. These long-term programs were aimed at innovative capabilities: Provide Real-time Total System Health Status; Predict Any System's Mission Capability; Proactively Maintain Readiness; Design for Integrated System Life Cycle Management & Intrinsic Reliability; and Autonomously Reconfigure Systems for Any Damage Condition.[37]

The substantial investment in time and resources that went into the planning and implementation of FLTCs, along with the strategic shift to longer term investments, resulted in a decreased emphasis on many program areas including sustainment. TD individual programs and established ATDs made up the major part of the sustainment portfolio. The influence of the FLTC process was evident in that the focus of sustainment programs within the S&T plan was moving toward

[35]Leo J. Rose. April 2008. Air Force Research Laboratory's Focused Long Term Challenges. Air Force Research Lab, Eglin Air Force Base, Florida, Munitions Directorate.

[36]Ibid.

[37]USAF. The Air Force Science and Technology Plan Fiscal Year 2000. Available at http://www.wslfweb.org/docs/st_plan00final.pdf. Accessed May 5, 2011.

Condition Based Maintenance—a program that would "enable total state awareness" and "eliminate time-phased maintenance."[38] One of the TD programs that continued to support the sustainment area was the Materials and Manufacturing Directorate's Air Force Manufacturing Technology Program (ManTech). Some of the many successful ManTech programs directed at sustainment are documented in Box 5-2.

ManTech programs such as Lean Depot Repair (LDR); Aerial Multi-Axis Platform (AMP), aimed at the de-paint process in the depots; Enhanced MAUS (Mobile Automated Ultrasonic System) Inspection Capabilities, aimed at improving the MAUS' detection capabilities; and the upcoming HVM program were all directed at improving flow time and reducing cost at the ALCs.[39] Lean Sustainment was an earlier ManTech program that produced the foundation on which the lean programs, such as LDR, were built. By the end of this period, the ATC process within AFMC had gone into somewhat of a holding pattern, resulting in no new ATDs being commissioned. In its place, a new process had been established for coordinating the needs of the developers and users of technology intended for sustainment. This Sustainment Technology Process:

> . . . provides a systematic repeatable method for identifying sustainment needs starting with requirements generation, going through validation, execution, transition, and ending with implementation. This process creates a strategic partnership between the Science and Technology (S&T) provider and the sustainment and acquisition communities to address sustainment technical opportunities, solution planning and programming. While this process is primarily focused on technology improvements within AFMC Centers, it will also be an avenue for MAJCOMs to identify sustainment technology needs that are of concern to both.[40]

The process is shown in Figure 5-5. The process leads to a meeting of the Senior Sustainment Steering Committee, headed by the AFMC/A4; indications are that this process has not been achieving its goals. Although stating otherwise, the complex effort to bring all of these parties together was not designed to establish requirements and ensure resources for transition across the "valley of death." Rather, as indicated clearly in Figure 5-5, it was to provide "advocacy" to those organizations controlling the resources. Efforts by such organizations as ASIP to identify technology "needs" could find no clear path through which to translate those needs into funded "requirements."

---

[38]Kathy Stevens. "Air Force Sustainment Science & Technology." Presentation at the CTMA 2009 Symposium, March 30-April 2, 2009, Detroit, Michigan.

[39]Ibid.

[40]AFMC Guidance Memorandum 61-101, January 27, 2007.

BOX 5-2
ManTech Success Stories

Organization
The ManTech program is chartered by U.S.C. Title 10, section 2521 and implemented by DODD 4200.15. The purpose of the program is to enable a robust industrial base for affordable warfighter weapon systems, and provide advanced manufacturing capabilities to multiple weapon systems. The goals of the program are to reduce acquisition and sustainment costs; reduce cycle time for technology transition, manufacturing, and repair; and improve quality, productivity, and business practices. The program focuses investments on those beyond the normal risk for industry and system program offices. ManTech directly supports administration, Congressional, Secretary of Defense and Secretary of the Air Force industrial base policies. Within the Air Force, program management responsibility is assigned to Air Force Materiel Command (AFMC), with execution performed centrally by the Manufacturing Technology Division, Materials & Manufacturing Directorate, Air Force Research Laboratory (AFRL/MLM).

Requirements and Planning
Technical requirements, priorities, and investment planning are developed in partnership with AFRL, Air Force program executive officers (PEO), ALCs, major commands, and industry. Technology transition requirements are identified and developed in partnership
with AFRL. Acquisition and sustainment requirements are identified and developed through integrated product teams in each customer "sector": aeronautical; sustainment; armament; directed energy; command, control, intelligence, surveillance, and reconnaissance; and space. Key criteria for investment decision are warfighter capability
impacts, pervasive system applications, and stakeholder implementation commitment. Requirements and plans are coordinated and approved by HQ AFRL, HQ AFMC, and SAF/AQR.

C-17 Landing Gear Doors
An increase in failure rates for C-17 main landing gear doors has become a major contributor to reduced mission capability of the aircraft. The AFRL Materials and Manufacturing Directorate and Air Vehicles Directorate, in cooperation with The Boeing Company, have successfully developed and implemented a durable composite C-17 main landing gear door that resolves the C-17s number one airframe maintenance problem, saves more than $6 million in life cycle costs, and increases mission readiness days by 90 per year. A result of the Composites Affordability Initiative (CAI) C-17 technology transition demonstration program, the new main landing gear doors incorporate several advanced manufacturing technologies and an improved design that will increase reliability by 40 percent.

Turbine Engine Components
Historically, methods for predicting the life of gas turbine engine rotor components have resulted in a conservative estimate of useful life. The "retirement for cause" program developed, integrated, and deployed advanced inspection and life-estimating technologies that are in use today at Oklahoma City Air Logistics Center permitting longer service life for many safety-critical, high-value components on the F-15, F-16, B-1B, and B-2.

This program—a collaboration between Air Force ManTech, materials S&T, and logistics centers—is also credited with enhancing safety, increasing inspection throughput, and introducing damage tolerance and probabilistic concepts-of-life management for engines. The program has realized more than $1 billion in cost savings. A related ManTech effort, Engine Rotor Life Extension, is currently working to create and implement the technology to further extend the life of these components and address the more complex rotor geometries of advanced engines for the F/A-22 and F-35, for an additional projected $550 million in savings.

Lean Depot Repair

The Lead Depot Repair project was established to determine whether lean methods could be adapted to the depot repair environment. Air Force ManTech partnered with Warner-Robins Air Logistics Center to revolutionize the programmed depot maintenance lines for F-15 and C-5 aircraft, reducing cycle time and cost. As a result, critical warfighting assets are more quickly moved from maintenance to fully operational status. Specifically, on-time return of C-5 aircraft increased from 25 percent in FY 2000 to 100 percent in FY 2004. The average flow time for depot maintenance on C-5s at the Air Mobility Command was reduced from 339 days in FY 2000 to an average of 240 days in FY 2004. On-time return of F-15 aircraft increased from 12 percent to 80 percent between FY 2000 to 2004. As a result, 22 additional F-15s (the equivalent of an additional squadron) were freed for operational use. Due to the success of this effort, lean depot practices are rapidly expanding across DOD, including Army arsenals and depots, Navy air depots, and shipyards.

SOURCE: Extracted verbatim from *Defense Science Board Task Force on The Manufacturing Technology Program: A Key to Affordably Equipping the Future Force.* February 2006.

Summary of Recent History

The past decade has seen significant accomplishments in the development and transition of technologies that have increased aircraft availability and reduced sustainment costs. During the same period, the Air Force S&T system has seen many changes in strategy and processes, each one initiating some new programs and disrupting others, leading to less than optimum use of limited resources. Unfortunately, at the end of this period, the process to transition sustainment technology into funded requirements was broken.

> **Finding 5-5.** The Air Force Materiel Command has a strong cadre of highly qualified researchers in AFRL and applied engineering talent in the ALCs. Historically, AFMC has made major contributions to the development and transition of technology in support of sustainment.

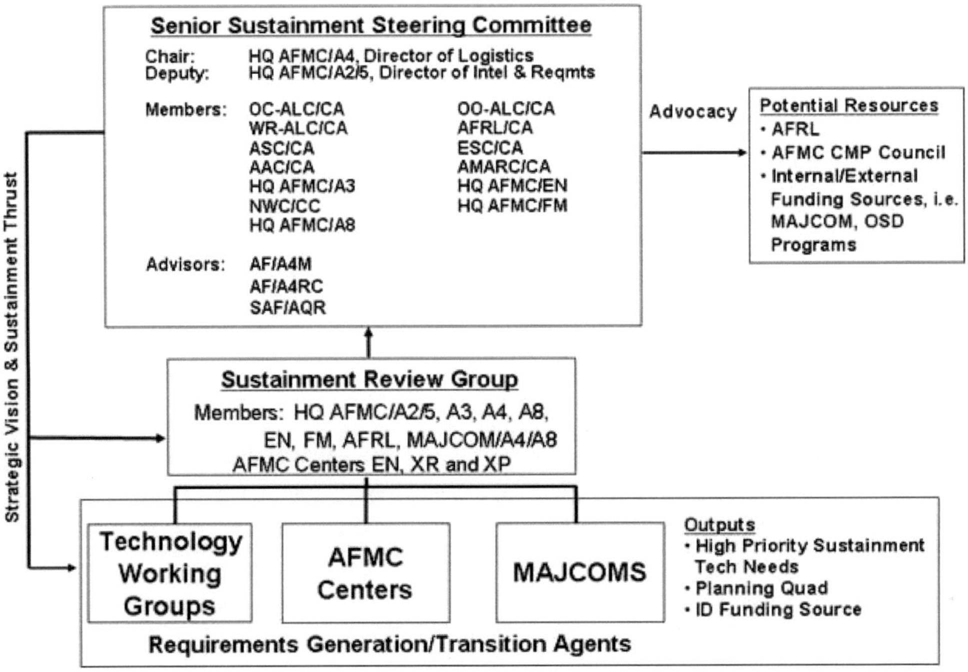

FIGURE 5-5
Air Force Materiel Command (AFMC) sustainment technology process. SOURCE: AFMC Guidance Memorandum 61-101, January 27, 2007.

Finding 5-6. In recent years, because of (1) frequent changes in management; (2) weakened processes for establishing requirements; and (3) reduced availability of resources, the Air Force Material Command talent base has not been optimally engaged in finding solutions to critical cost, schedule, or performance sustainment issues.

TRANSITION TO THE FUTURE

The Air Force appears to be addressing the severe criticisms of its S&T process.[41] It has done so in the context of rising sustainment costs that refocus attention on the potential for investment in technology to effectively increase aircraft availability and/or reduce sustainment costs. Addressing these challenges in parallel, the

[41]Steven H. Walker, Deputy Assistant Secretary of the Air Force for Science, Technology, and Engineering. "Air Force Science and Technology Strategy." Presentation to the Air Force Studies Board, November 16, 2010.

Air Force has adopted a strategy of ILCM and instituted new processes to manage the S&T portfolio. The following section discusses the current state of this transition under the headings of the three R's: requirements, resources, and right people.

Requirements

Guidance for establishing requirements in sustainment technology development are now to be drawn from the *Air Force Science and Technology Strategy 2010*, described earlier in this chapter. Sustainment, one of eight areas targeted for increased emphasis, will be planned along with other areas according to the approach sketched in that document (see Figure 5-6). This oversimplified diagram describes the development of technology as a linear feed-forward process, neglecting the important feed-back of information and experience that often energizes early-stage development. Nevertheless, it is a convenient planning structure.

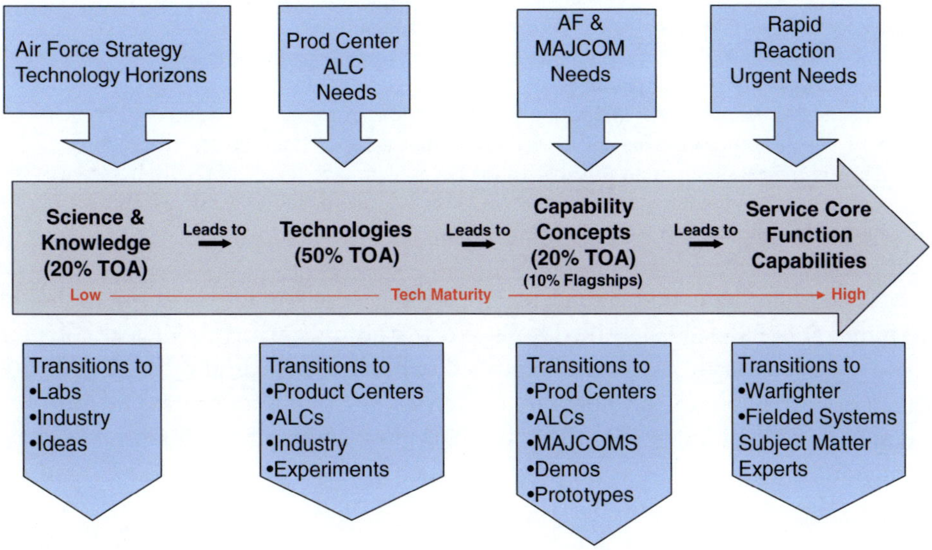

FIGURE 5-6
Air Force Research Laboratory (AFRL) approach to S&T. SOURCE: General Norton A. Schwartz, Chief of Staff, United States Air Force, and Michael B. Donley, Secretary of the Air Force, " Air Force Science and Technology Strategy 2010."

Quoting *Air Force Science and Technology Strategy 2010*, the four stages in this process are[42]

> Science and Knowledge
> Science and knowledge are the foundation of the Air Force S&T Program and the cornerstone of the future force. Based on visions of the future established by Air Force leadership, Air Force scientists and engineers identify, nurture, and harvest the best basic research to transform leading-edge scientific discoveries into new technologies with substantial military potential. These technologies transform the art-of-the-possible into the near-state-of-the-art and offer new and better ways for the acquisition community to address far-term warfighter needs.
>
> Technologies
> Air Force scientists and engineers continually interact with warfighters to understand their capability needs. The Air Force S&T Program addresses these needs by leading and harnessing innovation across service laboratories, government agencies, industry, and academia. These efforts mitigate risk and create the foundation for new capability concepts.
>
> Capability Concepts
> Senior representatives from Headquarters Air Force, MAJCOMs, Centers, and Air Force Research Laboratory (AFRL) will work together to define a balanced set of capability concepts that support known warfighter needs and mitigate risk from emerging threats. The highest-priority capability concepts are designated as Air Force "Flagship Capability Concepts (FCCs)." These FCCs address validated capability gaps and increase Air Force leadership's visibility into the Air Force S&T Program.
>
> Service Core Function Capabilities
> The Air Force's investment in S&T ensures the infusion of revolutionary and evolutionary S&T-enabled capabilities that are needed to maintain air, space, and cyberspace dominance. The Air Force S&T Program will address the needs identified in each of the twelve Service Core Functions (SCFs). Each of the MAJCOMS has one or more SCFs. Sustainment is housed in Agile Combat Support, AFMC's SCF.

As noted earlier in this chapter, technology development for sustainment should be found at every stage identified in Figure 5-6, but it is primarily in Technologies that transitions to the ALCs are identified, while both ALCs and MAJCOMs are called out as recipients in Capability Concepts. The planning process for sustainment within Technologies is primarily the responsibility of AFRL, demands close cooperation with the ALCs and programs such as ASIP, ENSIP, and FSIP in identifying priorities, and is endorsed by higher-level validation within AFMC. As this report is being written, this planning process has not yet been completed, although extensive pre-planning is evident. During 2010, AFRL commissioned a study by a

[42]General Norton A. Schwartz, Chief of Staff, United States Air Force, and The Honorable Michael B. Donley, Secretary of the Air Force. Air Force Science and Technology Strategy 2010.

distinguished group, seeking its advice on priorities and process.[43] During 2011, AFRL will be further aided by the results of the AFSAB study on sustainment that is under way in parallel to this current study and mentioned earlier in the chapter. Anticipating a successful planning process, the Air Force has already committed to an increase in funding for this area. This planned increase of funds is consistent with statements by the Greybeards that the present effort is underfunded, but until the planning process is complete, no judgment can be made about its adequacy to address the need. Also uncertain at this time are the specific process changes that may accompany this technology development plan and how priorities and funding will be made available for transition to ALCs and industry.

The planning that leads to Capability Concepts supporting the MAJCOMs is carried out as part of AFRL's Integrated Planning and Programming (IPP) process.[44] This "five body" integrated process, involving the TDs, a Capabilities Council, a Capabilities Working Group (CWG), an IPP Council, and the Commander, was developed to ensure a balanced S&T investment portfolio that addresses near- to far-term warfighter needs. The CWGs are responsible for managing customer-AFRL interface, gathering needs, and translating those needs into S&T projects that will deliver the desired capabilities. There is a CWG for each MAJCOM and its associated SCFs, and each CWG is chaired by a senior leader in the AFRL. In the case of sustainment, the AFMC CWG addresses AFMC's SCF of Agile Combat Support, which includes sustainment needs along with those of several other areas. This CWG is chaired by the Director of the Materials and Manufacturing Directorate of AFRL. Two organizational paths now exist for high-level validation and transition of sustainment technology development within AFRL: (1) selection at the Air Force Requirements Oversight Council (AFROC) in which a very few Capability Concepts are identified as Air Force Flagship Capability Concepts (FCCs) and, more commonly, (2) identification as Advanced Technology Development (ATD) programs through the Advanced Technology Council accessed via the Sustainment Technology Process.[45] The Capability Concepts/AFROC process identifies high-priority candidates that may be designated as FCCs. The definition, characteristics, and attributes of FCCs are shown in Figure 5-7.

Organized within AFRL, these candidate FCCs then follow the process shown in Figure 5-8 to become FCCs. In the first FCC submission under this new strategic management approach (completed in November 2010), AFRL submitted several

[43]Vince Russo. "Greybeard Assessment of the Sustainment Technology Transition Process." Presentation to the committee, Dayton, Ohio, February 7, 2011.

[44]Personal communication between C. Browning and Dr. James Malas, AFRL/XP.

[45]USAF. 2011. Sustainment Technology Process. Personal communication from Claudia Kropas-Hughes, Deputy Chief, Technology Transition Division, AFMC/A5S, to the committee, May 4, 2011.

- **Definition: An integrated technology project collaboratively developed by MAJCOM(s), Center(s), and AFRL that:**
  - **Addresses a documented and prioritized MAJCOM capability need**
  - **Is commissioned via Air Force S&T governance structure**
  - **Is traced to a CRRA gap and linked to a Service Core Function Master Plan**
- **Attributes:**
  - **Initial systems engineering and Development Planning (DP) initiated**
  - **Between a leading DP concept and a prototype**
  - **Assigned to lead Center for transition**
  - **MAJCOM transition manager identified**
    - **Transition funding (6.4) committed 2 years prior to S&T completion**
  - **Defined S&T baseline/exit criteria**
  - **S&T project ideally completed during current Future Years Defense Program**

1

FIGURE 5-7
Definition and attributes of Flagship Capability Concepts (FCCs). SOURCE: Steven H. Walker, Deputy Assistant Secretary of the Air Force for Science, Technology and Engineering. "AF Science and Technology Strategy." Presentation to the Air Force Studies Board, November 16, 2010.

FCC candidates, and the Air Force S&T Board recommended that the following projects be selected as FCCs:[46]

1. High Velocity Penetrating Weapon (HVPM)
2. Responsive Reusable Boost for Space Access (RBS)
3. Selective Cyber Operations for Tech Integration (SCOTI)
4. Low Observable (LO) Maintainability
5. Adaptive Versatile Engine Technology (ADVENT)
6. Selectable Effects Munitions (SEM)
7. Next Gen C2 and Operations for remotely Piloted Aircraft (RPA)

From this list, the AFROC selected the top three candidates.[47] The sustainment candidate, LO Maintainability, did not make the cut during this cycle. However it was characterized as an excellent program and will continue to support its prime

[46]Major General Ellen Pawlikowski. "AFRL Overview to the Electronic Engineering Steering Group." January 2011.

[47]Ibid.

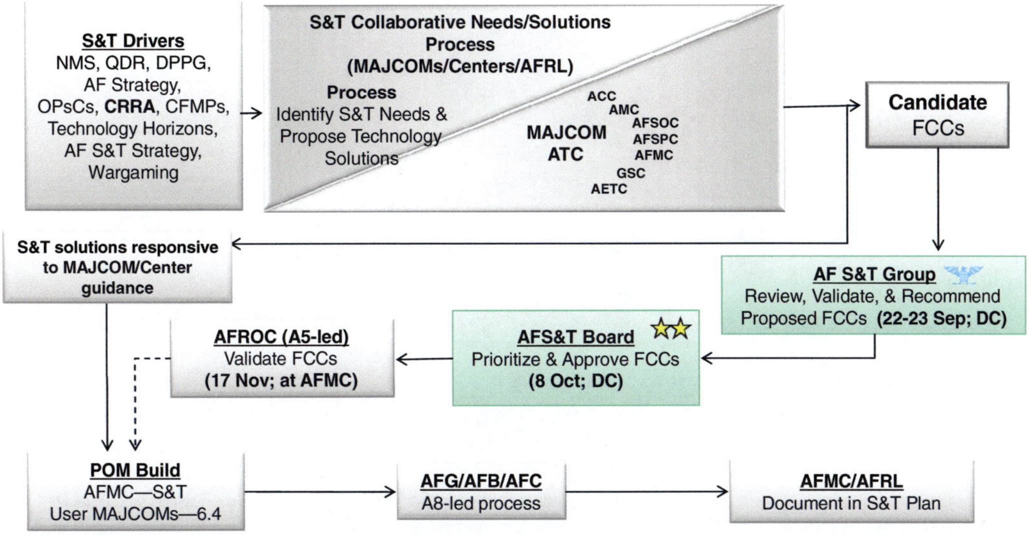

FIGURE 5-8
S&T planning process producing Flagship Capability Concepts (FCCs). SOURCE: Steven H. Walker, Deputy Assistant Secretary of the Air Force for Science, Technology, and Engineering. "AF Science and Technology Strategy." Presentation to the Air Force Studies Board, November 16, 2010.

customer, ACC. The other candidate programs will likewise continue to be worked by AFRL and its partners as Capability Concepts. This is certainly the highest visibility given to a sustainment technology program in recent memory and reflects well on the Air Force's commitment to reemphasize this important area. The second and more common transition path for sustainment technologies will lead to the AFMC/ATC via the Sustainment Technology Process. This process, identified earlier in Figure 5-5, has now been revisited and converted into the seven-step process depicted in Figure 5-9.[48]

This process was roughly at step 4 when this report was being written and is expected to lead to recommendation to the AFMC/ATC in 2012. In lieu of completion of the process this year, AFMC/A4 has identified HVM as the sustainment candidate for consideration by the AFMC/ATC scheduled for August 2011.

[48]USAF. 2011. Sustainment Technology Process. Personal communication from Claudia Kropas-Hughes, Deputy Chief, Technology Transition Division, AFMC/A5S, to the committee, May 4, 2011.

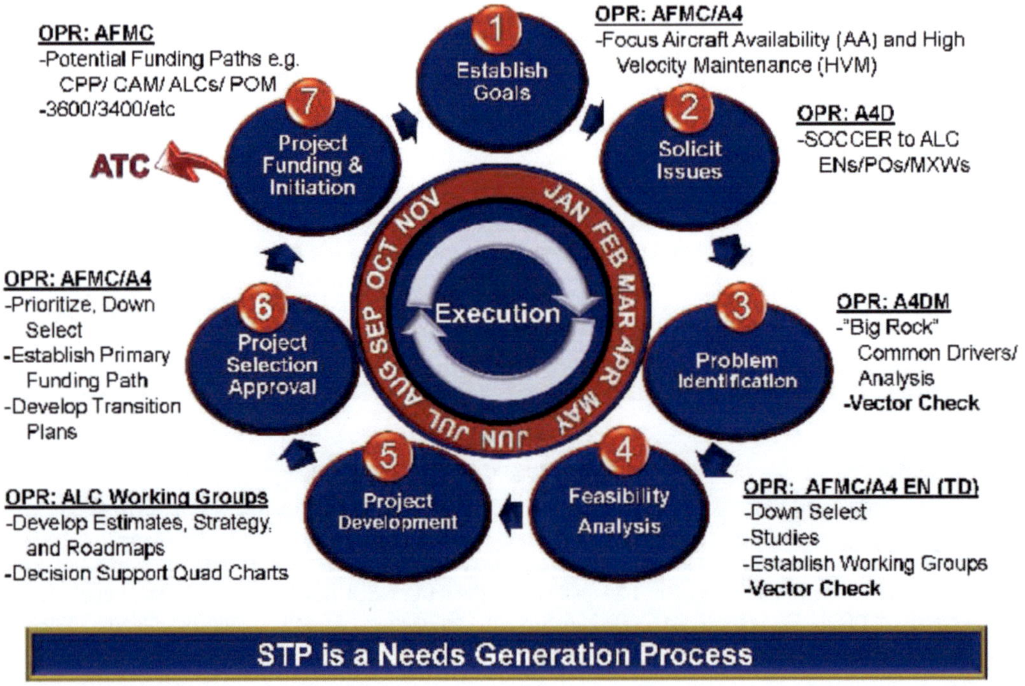

FIGURE 5-9
Newly revised sustainment technology process. SOURCE: Claudia Kropas-Hughes, Deputy Chief, Technology Transition Division, AFMC/A5S, personal communication to the committee, May 4, 2011.

Resources

Adequate and timely funding are key factors in any successful technology transition process. Funding issues affecting the sustainment technology transition process are (1) the restrictions placed on use of funds associated with each category of the "technology for sustainment" process: development, implementation, and maintenance; (2) the amount of funds in these categories required to support the work needed; and (3) the timing of the available funds. Funds are broken out by appropriations and are often referred to as "colors of money." There are several colors of money within the Air Force, but three are strongly tied to the "technology for sustainment" processes. Figure 5-10 illustrates the funding categories for the Air Force S&T program, along with those of the acquisition and operation and maintenance (O&M) programs, in addition to the various colors of money that house these program elements.

Their specified functions prohibit use of funds in any category for reasons other than those allowed by statutes. If funding for any stage in the technology transition shown in Figure 5-6 is inadequate, then it is difficult to cover the deficiency with funds from another color or level.

In addition to category issues, the successful transition of technologies for sustainment is greatly affected by the amount of funding within each of the stakeholders—the blocks within the levels shown in Figure 5-10—and the timing of these funds. Amounts and timing are firmly linked. Adequate funding at the S&T stage is essential to creating and sustaining technical expertise, producing a suite of technologies with potential sustainment applications, and developing totally new technology solutions to sustainment issues. Inserting S&T solutions that are not quick reaction support can be very time sensitive. In many cases, there is a window of opportunity where the need, S&T solution, system lifecycle stage, and funding all align. If the funding is not there, on time, this delicate balance can be disrupted, and the window can close very quickly. Because of the time-sensitive nature of the

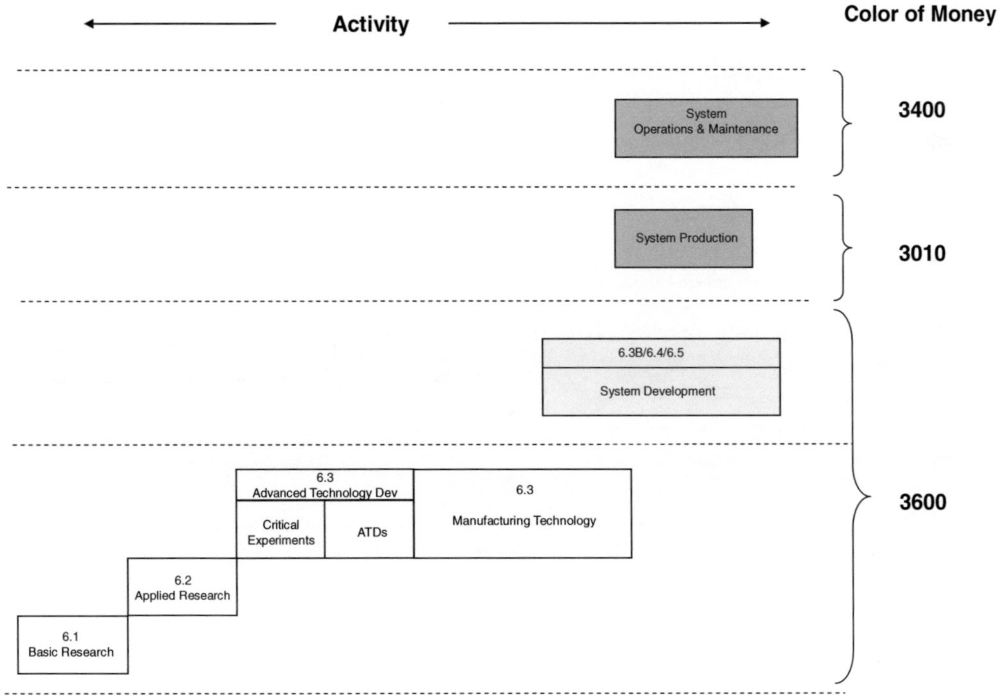

FIGURE 5-10
Program elements by funding levels.

insertion process, reprogramming of funds or developing new sources of funds may not be an option.

As shown in Figure 5-11, the overall DoD S&T budget requests for 6.3 funds have been relatively constant since FY 2003 and fairly flat for 6.1 and 6.2 funds since FY 1998. Congressional interest items and external entities, such as DARPA, have provided additional funds, but these are generally earmarked to specific organizations for specific technologies. Since the "cost of doing business," which includes salaries, contractor costs, facility charges, and supplies, has increased over this period, the net buying power of the S&T program has effectively decreased. The combination of AFRL's reduction in buying power with increased and changing warfighter needs has inevitably resulted in the prioritization of programs, with some going unfunded or moved to the out years. With the ever-present competition for resources, sustainment-related technologies require strong support from the highest levels to maintain its share of the Table of Allowance (TOA). As this high-level support waned so did funding for sustainment.

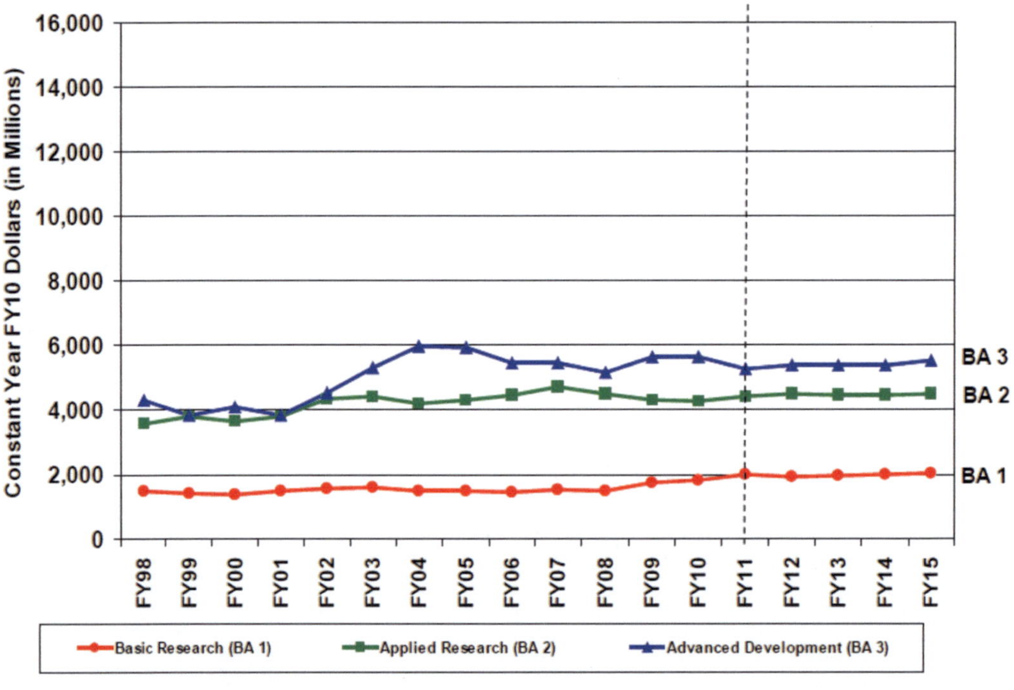

FIGURE 5-11
DoD S&T funding by budget activity. SOURCE: Bob Baker, Deputy Director for Plans and Programs, Office of the Director of Defense Research and Engineering. "Fiscal year 2011 President's Budget Request for the DoD Science & Technology Program." April 13, 2010. Available at http://www.dtic.mil/ndia/2010SET/Baker.pdf. Accessed on August 15, 2011.

Finding 5-7. Successful development and transition of sustainment technology require multi-year coordinated planning by several related organizations within the Air Force and with its suppliers. This planning must be organized and then validated at decision levels above those within the AFRL.

Right People

Qualified people at all levels of technical skill and training are critical to the health of any technology-dependent system. The Air Force is not alone in recognizing that weaknesses in its technical workforce are present and becoming more significant. Major national studies, including the NRC report *Rising Above the Gathering Storm,* have identified serious deficiencies in the current U.S. education system for science, technology, engineering, and mathematics (STEM) and have made recommendations for improvement at all education levels.[49] The DoD and individual services have all recognized this need and have instituted programs for STEM development of various kinds ranging from K-12 through to technical and graduate study and on to continuing education.

In the spring of 2011 the Air Force released *Bright Horizons,* a workforce strategic plan designed to support the S&T visions identified in the 2010 *Technology Horizons.*[50,51] Bright Horizons defines a process of strategic workforce management, identifies specific goals, and articulates a broad array of activities including identifying technical skill needs throughout the Air Force, encouraging continuing education opportunities, supporting undergraduate and graduate education, and engaging in K-12 outreach. Overarching responsibility for execution will be monitored by the newly created STEM Advisory Council, chaired by SAF/AQ. Although broad in scope and vision, this document is, by its very nature, short on specifics, including implications of such issues as budget and hiring freezes in the era of constrained budgets that lies ahead. It will be several years before sufficient detail is available to see how *Bright Horizons* influences the scientific and engineering workforce within the Air Force.

The committee did not perform sufficient research on the specific workforce needs in the areas covered by sustainment technology to justify detailed findings and recommendations. Somewhat unique to the case of sustainment, required expertise has historically been internally developed to be able to authoritatively supply quick reaction support, to serve as collocated engineers in the SPOs or

[49]NRC. 2007. Rising Above the Gathering Storm: Energizing and Employing America for a Brighter Economic Future. Washington, D.C.: The National Academies Press. Available at http://www.nap.edu/catalog.php?record_id=11463.

[50]USAF. 2011. Bright Horizons...the AF STEM Workforce Strategic Roadmap. Washington, D.C.

[51]USAF. 2010. Technology Horizons: A Vision for Air Force Science and Technology During 2010-2030 (Volume I). Washington, D.C.: Office of the Chief Scientist of the Air Force. May 1.

ALCs, and to understand the ALC/sustainment environment. It became clear during the study that the AFRL's commitment to developing and sustaining this unique talent pool appears to have gradually eroded over the past several years as sustainment has played a reduced role in Air Force S&T.[52] Recent S&T workforce development efforts have been directed at innovative, longer range technologies, rather than nearer term "support" technologies. Sustainment as a career field seems to have lost its luster within AFRL. Any plan developed by the Air Force to address its sustainment needs must necessarily be cognizant of this workforce issue. *Bright Horizons* creates the structure that would encourage identification of these critical needs and opportunities for addressing them.

RECOMMENDATION

AFRL can point to a long history of attention to sustainment and to many successful transitions to industry and the ALCs of technology that increased aircraft availability and/or reduced maintenance costs. In recent years, support for sustainment-focused technology has waned because of increased attention to other technical priorities and opportunities, too frequent changes in strategy and process, and reductions in funding for sustainment technology, especially for the transition process. There is capability within AFRL development programs to affect sustainment costs on existing weapon systems and to justify increased AFRL investments and foci in these areas. New processes, commitment to ILCM by management, and increased attention to sustainment at the highest levels have set the stage for optimum use of these resources. Missing is the comprehensive plan and subsequent implementation.

> **Recommendation 5-1.** The Air Force should develop a "technology for sustainment" plan that identifies processes, technical agendas, workforce needs, and required funding resources. Such a plan should be imbedded within the overall ILCM strategy that is being developed by the Air Force.

CONCLUDING THOUGHTS

Throughout its history, the Air Force S&T system has made numerous noteworthy transitions of technology, including intellectual capital, to the sustainment community, which includes the acquisition, industry, ALCs, and fielded systems. From emerging technologies to rapid reaction to urgent needs, AFRL personnel have delivered technologies to support the Air Force mission. For the most part, AFRL sustainment technology development has focused on the near- to mid-term

[52] Committee discussions with AFRL representatives on February 7-8, 2011.

timeframe. The emphasis on sustainment has ebbed and flowed over the years: it was a major emphasis area in the early years, was less so during the FLTC era, and is receiving increased planning emphasis in today's developing strategy. It is imperative in this changing environment that a comprehensive technology development and transition plan be completed and *implemented* with adequate resources and personnel to achieve the stated Air Force goals of ILCM.

During the past decade, a series of changes in processes for establishing requirements and allocating adequate resources has led to a far less than optimum usage of the highly qualified personnel in AFRL and the ALCs charged with sustainment tasks. This report was written during a period in which high-level changes to the S&T system were being developed but had not yet fully reached down into either the laboratory or the centers. The committee was encouraged to learn of the recommitment to a process that would lead to funding of high-priority areas to ensure transition. It was also encouraged to learn that sustainment is clearly identified in the recently released Air Force S&T plan and that a new program element has been established for this area. Pending rapid development of these new processes and their subsequent implementation, these steps all tend in the right direction for the Air Force.

Underlying all of the above is the issue of what specific technology development areas should be included in the mix when technology for sustainment is identified. Historically, sustainment technology referred primarily to issues developing during the life of already acquired systems. Detection of problems and technology for repair dominate that arena. Interaction with the ALCs and SPOs has been and remains critical to the identification of problems needing fixing and viable approaches to doing so. On the other hand, a broad area of technology development is intended to lead to longer life and less expensive maintenance that may be introduced into new systems. As the Air Force fully embraces the concept of ILCM now beginning to appear in high-level plans and visions, it will be necessary to broaden the common understanding of technology for sustainment to include those technologies and adequately support their development and transition into new systems. This subject is further explored in Chapter 6 under the heading *Providing for Continued Incorporation of Technology for Sustainment.*

6

Incorporating Sustainability into Future Designs

INTRODUCTION

This chapter addresses element 5 of the Terms of Reference (TOR), that is, "Identify and make recommendations regarding incorporating sustainability into future aircraft designs." More specifically, this chapter focuses on the importance of the involvement of all Air Force stakeholders in the sustainment process during each of the three major phases of a weapon system's life cycle: the concept and initial planning phase; the system design and development (SDD) phase, which includes development and implementation of the manufacturing processes; and the deployment and support phase.

The sections below discuss the following topics: (1) incorporating sustainability in the concept and initial planning, SDD, and deployment and support phases; (2) incorporating desirable design features and applying lessons learned during weapon system life-cycle phases; (3) owning data rights/access and the ability to gain weapon system sustainment domain knowledge; (4) considering a blended-support concept; (5) moving to a data-driven sustainment strategy and common enterprise management in new designs; (6) providing for continued incorporation of technology for sustainment; (7) adopting the unique sustainment aspects with respect to rapidly fielded systems; and (8) understanding commercial aviation practices for Air Force consideration.

As with other TOR elements, there is a degree of overlap between the subjects covered in this chapter and in the other chapters. In most cases, this chapter refers to, rather than repeats, details found in earlier chapters. The discussion on continued incorporation of software and technology in future systems builds on the earlier discussion found in Chapters 4 and 5.

INCORPORATING SUSTAINABILITY IN THE CONCEPT AND INITIAL PLANNING, SYSTEM DESIGN AND DEVELOPMENT, AND DEPLOYMENT AND SUPPORT PHASES

Air Force personnel responsible for sustainment recognize the need to address sustainment in the life cycle of any weapon system prior to Milestone A to ensure that sustainable design attributes and support concepts are appropriately considered.[1,2,3,4,5,6,7]

Special and early emphasis must be placed on the critical nature of sustainment and on the importance of making decisions regarding the terms and conditions for data rights and the details of the manufacturing processes. Contractual arrangements between the Air Force and contractors regarding the support concept to be used also occur early in the life cycle. Sustainment professionals need to be involved early to influence the weapon system design and support concepts for sustainability.

Early involvement of sustainment personnel, both at the unit and depot level, is important because modern weapon systems can require long-term planning to provide the special skills training, transition original equipment manufacturer (OEM) repair capability, acquire security clearances for depot personnel, and establish a quality maintenance work force as well as to give feedback to the contractor about what works and does not work in a weapon system. In addition, early planning for sustainment will help the Air Force to efficiently meet the legislated guidelines for depot workload split that require long-term planning and budgeting for new

[1]Debra K. Tune, Principal Deputy Assistant Secretary of the Air Force for Installations, Environment and Logistics, Office of the Assistant Secretary of the Air Force for Installations, Environment and Logistics. "Developing the Right Product Support Concepts for the Future." Presentation to the committee, October 20, 2010.

[2]Findings and recommendations related to incorporating requirements into policy and implementing policy into effective practice are presented in Chapter 2.

[3]Blaise J. Durante, Deputy Assistant Secretary for Acquisition Integration, Office of the Assistant Secretary of the Air Force for Acquisition, "Budgeting Considerations Related to Sustainment." Presentation to the committee, October 21, 2010.

[4]Major General Kathleen D. Close, Director, Logistics and Sustainment, Air Force Materiel Command. "Weapons System Sustainment." Presentation to the committee, December 8, 2010.

[5]General Donald J. Hoffman, Commander, Air Force Materiel Command, Wright-Patterson Air Force Base. Personal remarks to the committee, December 9, 2010.

[6]Warner-Robins Air Logistics Center (WR-ALC) leadership and program managers. Personal conversations with the committee, WR-ALC site visit, January 5-6, 2011.

[7]Oklahoma City Air Logistics Center (OC-ALC) leadership and program managers. Personal conversations with the committee, OC-ALC site visit, January 11-12, 2011.

facilities and special equipment. Early involvement of sustainment professionals is also important for vetting technologies that will affect sustainability. Particular emphasis was placed on the early incorporation of lessons learned from recent experience in software, simulation, and low observables maintenance into future weapon systems planning early.

A recurring theme from sustainment professionals is that they should be involved in the life cycle prior to Milestone A of programs. Current sustainment requirements, as defined in Department of Defense (DoD) 5000 series instructions and AFI 63-101, are sufficient to define the process for ensuring sustainment decisions are made at the appropriate phase of the life cycle. However, a review of the development history of recent programs revealed that the sustainment community was not appropriately considered throughout the development process.[8] Congress recognized the need for the sustainment community to be involved in the early phases of the acquisition process.[9]

With enactment of the 2009 Weapon System Acquisition Reform Act (WSARA), the Secretary of Defense released guidance that increased the emphasis on sustainment and total lifecycle management in major weapon system programs. The FY2010 National Defense Authorization Act, Section 805, established the Product Support Manager (PSM) position for all Acquisition Category Acquisition Category (ACAT) I and II programs. Requirements were outlined in DTM 10-015, "Requirements for Life Cycle Management and Product Support." DTM-015 lists the following tasks:[10]

A. Provide weapon systems product support subject matter expertise to the PSM for the execution of the PSM's duties as the Total Life Cycle Systems Manager, in accordance with DoD Directive 5000.01.

---

[8]The reasons for this can be traced back to two changes made in the early 1990s. First, the 1994 Federal Acquisition Streamlining Act (FASA) stressed the use of commercial items and performance-based acquisition strategies. In particular, the FASA significantly reduced the number of military standards and specifications in favor of industry standards, and allowed industry to manage its own configuration data and use its own data systems until the end of the development phase. Second, the accompanying Air Force acquisition workforce reduction, which eroded the organic capabilities for acquisition management of development, logistics, and sustainment, created too much reliance on the contractor serving as the lead system integrator.

[9]Peter Levine, Senate Armed Services Committee, General Counsel, Readiness and Management Support Subcommittee, Majority Lead; Lynn Williams, House Armed Services Committee, Readiness Subcommittee, Majority Lead; and Vickie Plunkett, House Armed Services Committee, Readiness Subcommittee, Minority Lead. Discussion with the committee, February 16, 2011.

[10]DoD. "Directive-Type Memorandum (DTM) 10-015 –Requirements for Life Cycle Management and Product Support." October 6, 2010. Washington, D.C.: Office of the Under Secretary of Defense. Available at https://acc.dau.mil/CommunityBrowser.aspx?id=399932. Accessed February 16, 2011.

B. Develop and implement a comprehensive, outcome-based, product support strategy.
C. Promote opportunities to maximize competition while meeting the objective of best-value long-term outcomes to the warfighter.
D. Seek to leverage enterprise opportunities across programs and DoD Components.
E. Use appropriate analytical tools and conduct appropriate cost analyses, including cost-benefit analyses, as specified in Office of Management and Budget Circular A-94, to determine the preferred product support strategy.
F. Develop and implement appropriate product support arrangements.
G. Assess and adjust resource allocations and performance requirements for product support, not less than annually, to meet warfighter needs and optimize implementation of the product support strategy.
H. Document the product support strategy in the Life Cycle Sustainment Plan (LCSP).
I. Conduct periodic product support strategy reviews and revalidate the supporting business case analysis prior to each change in the product support strategy or every 5 years, whichever occurs first.

Air Force executives described the changing environment that recognizes earlier sustainment planning as codified by the WSARA. Figure 6-1 outlines how the Air Force views the process.[11]

Full implementation of new policy guidance may require staffing adjustments, and results will not be known for some time. However, the Air Force has energized its commitment to sustainment with the release of AFI 63-101 and the *Acquisition Sustainment Tool Kit (ASTK) Kneepad Checklist.* Revised Air Force strategy appears to favor a blended partnership between the Air Force and contractor capabilities across all areas of sustainment as shown in Figure 6-2.[12] As part of this strategy, the Air Force accepts some reliance on industry capabilities throughout the weapon system development and blended capabilities in the program's sustainment phases. However, for recent programs (e.g., C-17, F-22), it appears that overreliance on industry in the development phases has left the Air Force unable to easily stand up capabilities for organic maintenance because it lacks the data rights, domain

[11]Debra K. Tune, Principal Deputy Assistant Secretary of the Air Force for Installations, Environment and Logistics, Office of the Assistant Secretary of the Air Force for Installations, Environment and Logistics. "Developing the Right Product Support Concepts for the Future." Presentation to the committee, October 20, 2010.

[12]Ibid.

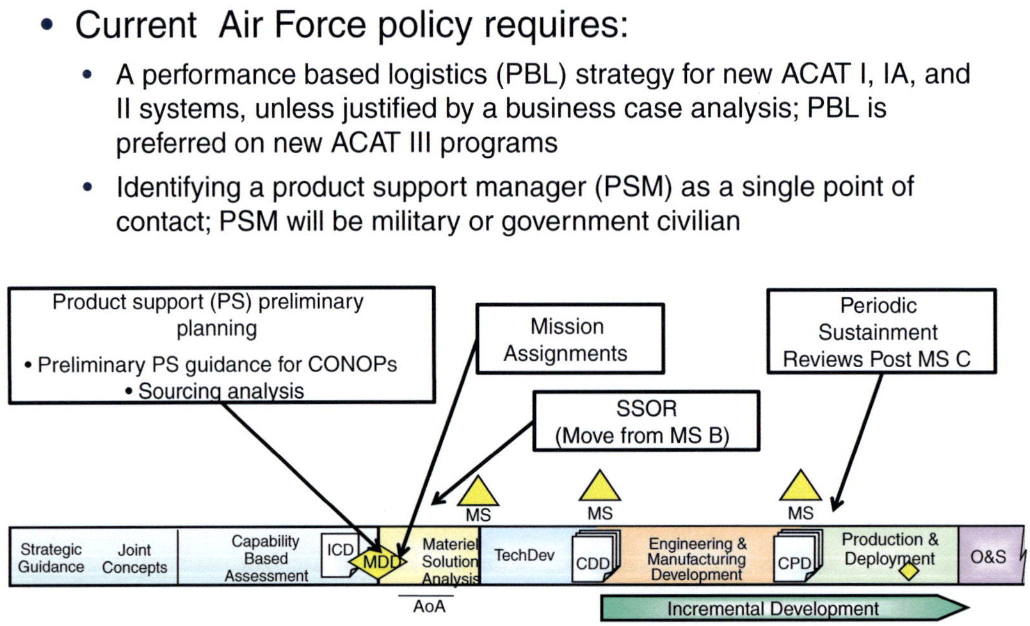

FIGURE 6-1
A changing environment. ACAT, Acquisition Category; AoA, analysis of alternatives; CDD, Capabilities Development Document; CPD, Capability Production Document; ICD, Initial Capabilities Document; MDD, Material Development Decision; MS, milestone; SSOR, Strategic Source of Repair. SOURCE: Debra K. Tune, Principal Deputy Assistant Secretary of the Air Force for Installations, Environment and Logistics, Office of the Assistant Secretary of the Air Force for Installations, Environment and Logistics. "Developing the Right Product Support Concepts for the Future." Presentation to the committee, October 20, 2010.

knowledge, and access to the tools required to manage the process.[13,14,15] Data rights, domain knowledge, and process management tools are addressed later in this chapter.

The development of a weapon system is a complex undertaking that results in numerous technical and manufacturing risks that drive program level re-planning and decision making. As the development proceeds and cost/schedule risks emerge, it is clear that recent programs have deferred the costs of establishing critical sustainability and training activities to allow completion of development and early

[13]Ibid.

[14]Ogden Air Logistics Center (OO-ALC) maintenance personnel. Personal communication with the committee, OO-OLC site visit, January 31-February 1, 2011.

[15]Grover L. Dunn, Director of Transformation, Deputy Chief of Staff for Logistics, Installations and Mission Support, Headquarters U.S. Air Force. "Expeditionary Logistics for the 21st Century (eLog21)." Presentation to the committee, January 17, 2011.

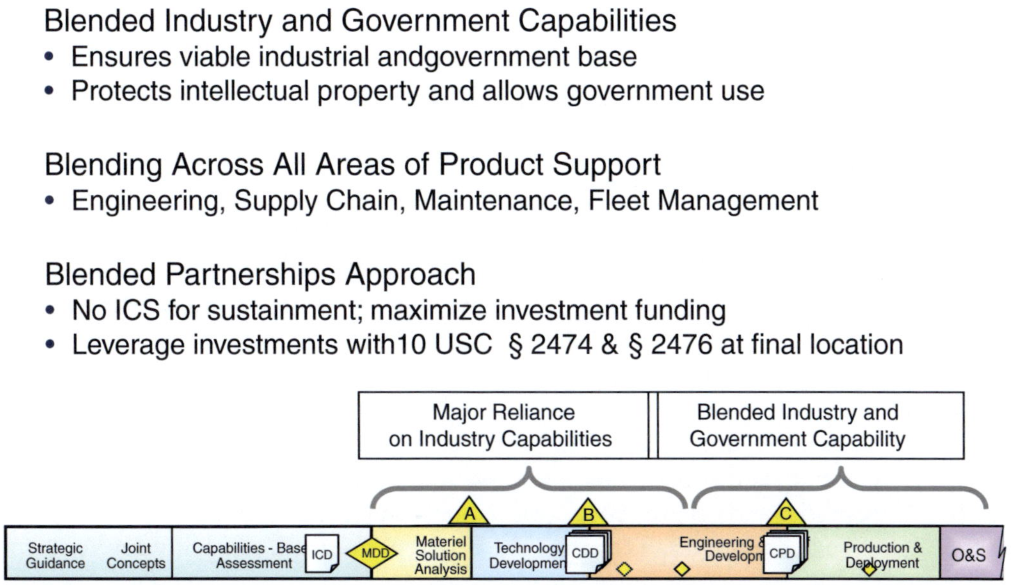

FIGURE 6-2
Evolving partnerships over the lifecycle of a weapon system. SOURCE: Debra K. Tune, Principal Deputy Assistant Secretary of the Air Force for Installations, Environment and Logistics, Office of the Assistant Secretary of the Air Force for Installations, Environment and Logistics "Developing the Right Product Support Concepts for the Future." Presentation to the committee, October 20, 2010.

production activities. These are prioritization decisions that can and should be made by the System Program Manager (SPM) and Procurement Executive Officer (PEO). However, the total cost of these activities, which must occur at some point in the lifecycle, must remain in program planning with visible estimated costs throughout at least the current planned system lifecycle, and the PSM should be actively involved and able to report such planning.

The WSARA and the Air Force implementation planning should allow greater influence on the sustainment process. Having sustainment professionals play an active role in the support system concept development ensures that sustainment requirements are properly introduced prior to Milestone A of the weapon system program. The opportunity now exists during SDD to ensure that the design incorporates the Integrated Life-Cycle Management (ILCM) perspective on support features and to initiate the development of technologies that assure the designing in of high component reliability, maintainability, and efficient repair techniques prior to system deployment.[16]

[16]Discussions with Air Force Research Laboratory (AFRL) personnel described in Chapter 5 provide insight into new materials and design features that should be addressed by the Air Logistics Centers

Finding 6-1. Sustainment professionals are not actively involved in the early Defense Acquisition Management System (DAMS) phases of weapon system development. As a result, sustainment needs, planning, and costs are not being fully captured.

Recommendation 6-1. The Air Force should involve its sustainment professionals throughout the DAMS, from Materiel Development Decision to Production and Deployment, in all future weapon systems development. Funding for sustainment planning and support should be given the same visibility as that for the development of capabilities and performance.

Recommendation 6-2. The Air Force should give funding for sustainment planning and support the same visibility as that given to development of capabilities and performance. Contractor sustainment activities should be developed and tracked as a contract line item separate from capability development.

INCORPORATING DESIRABLE DESIGN FEATURES AND APPLYING LESSONS LEARNED DURING WEAPON SYSTEM LIFE-CYCLE PHASES

Through telephone interviews with two senior engine company engineering executives [17] and visits with members of the ALCs, [18] the committee developed a list of desirable sustainment attributes (Box 6-1).

These discussions highlighted the importance of involving personnel with strong sustainability experience in the system design and development phases and extending into the manufacturing phase, where processes that may impact maintainability are implemented. The discussions with the engineering managers also noted the value of the "Blue Two" program of the early 1990s in which contractor engineers were integrated into Air Force field maintenance teams and were allowed to see first-hand some of the issues associated with maintaining complex equipment in operational environments. A Society of Automotive Engineers (SAE) paper

(ALCs) as new designs emerge. Existing programs, such as the Component Improvement Program (CIP), address issues that arise in fielded systems and help to accelerate maturation of new systems. SOURCE: Valerie Dahlem, Chief, Fighter Engine Programs Branch, Wright-Patterson Air Force Base. "Propulsion Sustainment." Presentation to the committee, December 7, 2010.

[17]Norm Egbert, Retired Vice President of Engineering, Rolls-Royce North America and Frank Gillette, Jr., Chief Engineer, F119 Project, Pratt & Whitney. Personal communication with the committee, April 6 and April 8, 2011.

[18]WR-ALC site visit by the committee, January 5-6, 2011; OC-ALC site visit by the committee, January 11-12, 2011; OO-ALC site visit by the committee, January 31-February 1, 2011.

BOX 6-1
Sustainment Attributes Checklist

1. Articulate a well-defined mission and sustainment plan that the design and development engineers can address.
2. Secure the commitment of both Air Force and contractor management teams to sustainment goals and requirement.
3. Directly involve Air Force System Program Office personnel as members of the design and development team.
4. Design the system to "robustness criteria" with over tests for vibration, temperature, and subsystem (sensors) failure with emphasis on both high- and low-cycle fatigue life.
5. Select durable and highly reliable components to reduce maintenance and replacement (extremely important in Low Observable [LO] platforms to avoid intrusion into LO materials areas). For engine components, involve subcontractors in maintainability and have components only one layer deep and removed with their related module.
6. Condition monitoring systems to provide insight into maintenance needs prior to scheduled or unscheduled maintenance actions.
7. Design systems to be modular and accessible to avoid unnecessary intrusion into components not requiring maintenance.
8. Locate components requiring high levels of either scheduled or unscheduled maintenance where they can be easily accessed.
9. Minimize (or eliminate if possible) the need for special tools.
10. Develop the inspection and repair techniques as part of the engineering and manufacturing development processes.
11. Provide maintenance instructions, training, and specifications in contemporary electronic format for application to depot activities to facilitate training, improve efficiency, and reduce the amount of engineering decisions on the floor.
12. Accept that software will evolve throughout the development process and into weapon system deployment.

about the F-119 engine program provides an excellent example of how sustainment considerations can be incorporated into design:

> One of the most important requirements for the F119 engine was that only five hand tools should be used to service the entire engine. All Line Replaceable Units (LRUs) would have to be "one-deep," meaning that the engine would have to be serviceable without removal of any other LRUs, and each LRU would have to be removable using a single tool within a 20-minute window. The most desired "design for sustainability" feature that was noted

> was to design for fewer depot visits in the systems life time and reduce the shop time per visit by design.[19]

A formal approach to using sustainment requirements in the design phase, called Human Systems Integration (HSI), was recently addressed by Liu and colleagues.

> *Human Systems Integration* (HSI) is defined as the "interdisciplinary technical and management processes for integrating human considerations within and across all system elements; an essential enabler to *systems engineering* practice" (Haskins, 2007). The primary objective of HSI is to integrate the human as a critical system element, regardless of whether humans in the system function as individuals, teams, or organizations. The discipline seeks to treat humans as equally important to system design as are other system elements, such as hardware and software.[20]

Committee members with experience in airline support noted that important lessons can be learned from commercial aircraft experience, and there was general agreement with this premise when it was raised in the telephone interviews and in a committee briefing by a senior airline maintenance manager. Examples were noted in cases where there was commonality between the commercial and military variants.

During its visits to the ALCs, the committee was briefed on the maintenance complexities introduced by the introduction of low observable (LO) technology in deployed systems. It is important for the design, development, and manufacturing functions for future designs to incorporate the lessons learned from the experience with these systems. Considerations include design for simplicity and durability with an emphasis on seals, panels, and edges designed for supportability. Doors and edges or access areas must be damage tolerant, and repairs must be modeled in representative environments such that the majority of unscheduled maintenance events can be accomplished without requiring LO restoration. A high premium was also placed on designing systems such that flight test or complex ground verification testing would not be required after LO maintenance. As outlined in the proceeding paragraphs, the Air Force has enjoyed isolated successes in capturing maintainability and sustainability improvements in weapon systems. However, the successes are not widespread, and opportunities for large-scale improvements have not been institutionalized.

---

[19] F.C. Gillette, Jr. 1994. Engine design for mechanics. SAE International. ISBN 1560915382. Warrendale, Pennsylvania: Society of Automotive Engineers.

[20] K. Liu et al. 2010. The F119 engine a success story of human system integration in acquisition. April 1. Defense A.R. Journal. Available at http://www.dau.mil/pubscats/PubsCats/AR%20Journal/arj54/Liu%2054.pdf. Accessed May 6, 2011. p. 286.

Finding 6-2. The Air Force does not have an institutionalized process for collecting and consistently incorporating desirable design features or applying lessons learned into the requirements documents for new systems or systems being modified. This results in frequent "re-invention" of improvements that could reduce costs and enhance the system supportability, maintainability, and availability.

Recommendation 6-3. The Air Force should establish an institutionalized process for collecting and consistently incorporating desirable design features or applying lessons learned from legacy programs into the requirements for new systems or systems being modified in the support phase of the life cycle and into the internal procedures involved in sustaining these systems.

DATA RIGHTS/ACCESS AND THE AIR FORCE'S ABILITY TO GAIN WEAPON SYSTEM SUSTAINMENT DOMAIN KNOWLEDGE

The importance of the Air Force acquiring adequate weapon system design data was stressed repeatedly to the committee. Current engineering design data must be maintained and delivered as a contract requirement to permit the systems to be supported organically when operationally deployed or in the later years. Most of this data is contained in digital form in electronic data exchange systems such as engineering product data management and software development environments. Future sustainment requires both access to the digital data and understanding of the computing environments in which they were developed. This was noted as being extremely important in cases when the mission and usage of the system changes from the original design intent. Several cases were noted in which the OEM had exited the business or in which the contractor did not view continued engineering support of the system to be economically attractive.

Even if the data are available, they can be difficult to use without knowledge of the engineering design standards and guidelines used to develop the data, which are often considered to be proprietary by the contractor. Participation by sustainment professionals in the design process allows for information and domain knowledge transfer beyond that which can be gained from deliverable information. Participation in the design process also helps to determine what data are best to procure in the development phase for future weapon system support.

Finding 6-3. The Air Force often does not have the required data rights/access and sufficient domain knowledge to facilitate easy transition to organic support of the overall supply chain and a number of technologies. This is exacerbated by industry proprietary design data, processes, and tools.

Recommendation 6-4. The Air Force should place more emphasis and implement additional training for acquisition professionals on the need for, the how to, and the pricing of proper data rights and domain knowledge related to the weapon system.

CONSIDERING A BLENDED SUPPORT CONCEPT

A blended support partnership between the Air Force and the contractor can provide an efficient support concept if instituted early in program planning. During the initial deployment and production phases of the program, the Air Force can realize cost efficiencies because of the contractor's greater knowledge of the aircraft and ability to combine production and sustainment supply chain activities. In the later stages of the aircraft life cycle, the resources and expertise within the Air Force's organic sustainment enterprise are invaluable. However, there have been instances in recent acquisitions of long-term performance-based logistics arrangements being established without consideration of the impacts to the Air Force sustainment enterprise.[21] As a result, the planning for conversion to organic support was inadequate, and the Air Force incurred unplanned facilitization, equipment, and manpower costs. The support issues include shutdown and storage/maintenance use of production manufacturing tools, stand-up of repair facilities and processes at the ALCs, adequate manpower planning, and transition of data/domain expertise from the contractor to the government.

Finding 6.4. The Air Force has not established an institutionalized process for ensuring that blended partnerships are given real consideration across the full lifecycle.

Recommendation 6-5. The Air Force should commit to establishing processes and resources that support consideration of a blended organic-contractor partnership early in the program life cycle and throughout the deployment and support phases.

[21]Debra K. Tune, Principal Deputy Assistant Secretary of the Air Force for Installations, Environment and Logistics, Office of the Assistant Secretary of the Air Force for Installations, Environment and Logistics. "Air Force Studies Board Sustainment Study: Developing the Right Product Support Concepts for the Future." Presentation to the committee, October 20, 2010.

MOVING TO A DATA-DRIVEN SUSTAINMENT STRATEGY AND COMMON ENTERPRISE MANAGEMENT IN NEW DESIGNS

The initiation of future aircraft designs provides an opportunity to incorporate not only configurations and features that will enhance the maintainability of the system, but also more efficient management tools and processes into the enterprise that will sustain the system. An effective data-driven sustainment strategy throughout the life of the weapon system was discussed in Chapter 2 and is recognized as an important element of any new system. The data and enterprise management systems for future aircraft programs should consist of four fundamental elements:

1. The design and development of an effective maintenance program.
2. An effective information system that tracks the process and how the equipment operates.
3. Continuous analysis and update of the requirements based on the information coming from the operational and maintenance actions.
4. An enterprise reporting and planning system that provides sustainment personnel the visibility and control over the processes needed to meet operational demand and requirements.

After reviewing the Air Force's programs and processes and how they are being implemented at the various ALCs, the committee concluded that some of the above elements are in place but it is not obvious that the first three are consistently performed or made available. To a great degree, a variety of efforts were neither uniformly performed nor applied. Each effort is useful in and of itself; however, the potential benefits cannot be fully realized if the efforts are not integrated.

The importance of information and data analysis to the improved effectiveness of sustainment of new designs cannot be underestimated. The following section addresses opportunities for migration to improved data management systems in support of the sustainment functions. Data and information systems must be integrated to produce a true view of overall sustainment operations as opposed to multiple independent sustainment operations. The organization that is responsible for much of the information analysis and management is often independent of the organization that performs the work and thus does not share common performance objectives.

Sustainment is a very complex process. Although many sustainment elements can be viewed individually, their integration into a holistic picture or pattern is essential. This integration is dependent on data integration. One key element of the sustainment process is maintenance.

Effective maintenance is as much about proper use of information as it is about actual accomplishment of the tasks. Only through a thorough analysis of the operational and maintenance data can the safety and efficiency benefits be fully realized. The four fundamental elements each have unique data requirements, which are key to developing the specific individual steps or tasks before all four elements interact with each other during the equipment's life cycle. It is a continuum that begins with an analytical approach then leads to adjustments and improvements based on service experience to change the design or maintenance program or modify the equipment.

The Data or Information Systems Needed to Develop the Maintenance Program

Development of the maintenance program should begin with the design process, well in advance of putting the equipment into operational status. The data are derived from the original engineering tests and analyses from the design phase of the aircraft, are based on Failure Mode Effect and Criticality Analysis (FMECA), and take advantage of service experience with similar designs and materials. Both the designers and maintainers must participate in the design process; they need each other's inputs and access to each other's information and experience. The process develops the actual tasks and categorizes them based on safety and operational implications. Although the safety and operational determination is often in the realm of the designers, the timing of task performance and the tasks' effectiveness is often in the realm of the maintainers. Reaching a common understanding based on information about what makes things work and what keeps things working leads to effective programs. This is a very important concept because neither the designer nor the maintainer will have the full knowledge and experience to fully develop a program in an individual vacuum.

The balance of performance considerations with service experience is the basis of program development. This first phase forms the foundation of effective sustainment for the life cycle of the equipment, and its associated data must be available throughout the service life of the equipment. In the commercial world this is accomplished through a cooperative effort between the manufacturer, regulators, and airlines using the Maintenance Steering Group Three (MSG-3) process techniques. During a visit to Warner Robins Air Logistics Center (WR-ALC), a committee member saw that the Air Force is adopting this approach but not uniformly across the total fleet.

Information and Data Needs Once a Platform Enters Service

Putting equipment into service requires a variety of data systems and information. Once the maintenance program is in hand, several critical information inputs are necessary: Where and how is the equipment going to operate? What facilities are available, and what facilities will be needed for the equipment type? What specialized skills will be needed? What levels of inventory and manpower will be needed? What tooling and equipment will be needed? These are just a few of the many questions whose answers have implications for information and data systems. To arrive at the answers, a database from which to draw previous experience and performance of similar and currently operating systems is needed. In the Air Force, a database (REMIS) is used by most weapon systems to track maintenance information. However, this database is a legacy system with wide variations in use.

An Enterprise Resource Planning (ERP) system is critical to the ability to match the planning needs to bring the people, parts, and equipment together at the right place and time. The Air Force is developing a common ERP system, but it is still a long time away. In the meantime there is a clear lack of effective planning and data collection systems. An ERP system is one of the foundations of sustainment. Developing an effective ERP system is quite difficult, and top-down/bottom-up involvement and stakeholder buy-in are essential. System development is not only about software but also about processes and behaviors. Lastly the ERP system is not a stand-alone system. It must be tied into an analysis process to determine that planning and implementation are effective and properly applied across the equipment and the facilities. The existing processes across the various ALCs are inconsistent, which can disguise weaknesses or failures in the process or produce misleading results. Thus, an effective loop of feedback and findings and organizational alignment is needed.

Information System Needed for Continuing Analysis and Surveillance of the Sustainment Process

When developing effective reliability and maintainability programs, it is important to possess good information about the operating performance of the equipment and the condition of the parts. The status of the maintenance program in terms of time and cycles is a basic but critical need of Conditioned-Based Maintenance. With this information, all of the inputs can be analyzed and results looped back into the maintenance program to continuously improve effectiveness or optimize operational results.

Some measure must be developed to determine how well the program works, how well the suppliers perform, and how well the workers achieve their goals. Obvious measures such as time on wing are essential, but tracking labor hours to determine economic benefits and cost at supplier on a unit basis is also important. Other measures such as delays caused by parts shortages, manpower shortages, waiting on approvals, and sign offs provide useful information. The use of these measures can only work if there is both informational and organizational alignment.

The Air Force has a matrix management organization. Matrix management can only work effectively if there is a transparent information system that measures what and who improves the system and what and who impedes the system. Unfortunately without effective and transparent systems, performance measurement is very difficult, and consequently the expected performance can deteriorate without designated responsible officials. The committee found that recent weapon systems, namely C-17 and F-22, have adopted data-driven strategies and enterprise management tools and are performing well above other weapon systems as a result. However, these systems have been developed as proprietary contractor systems with full contractor management and will not necessarily support future Air Force eLog21 and enterprise management concepts. Figure 6-3 shows the current proliferation of contractor-supported weapon system (CSWS) tools. To meet future sustainment requirements an integrated approach must be achieved.

Finding 6-5. The Air Force sustainment data collection effort uses a wide variety of legacy systems that are not optimized for enterprise information collection, analysis, or problem resolution.

Recommendation 6-6. The Air Force should develop future systems to support a common Air Force enterprise management system, eliminating proprietary contractor data management and exchange approaches.

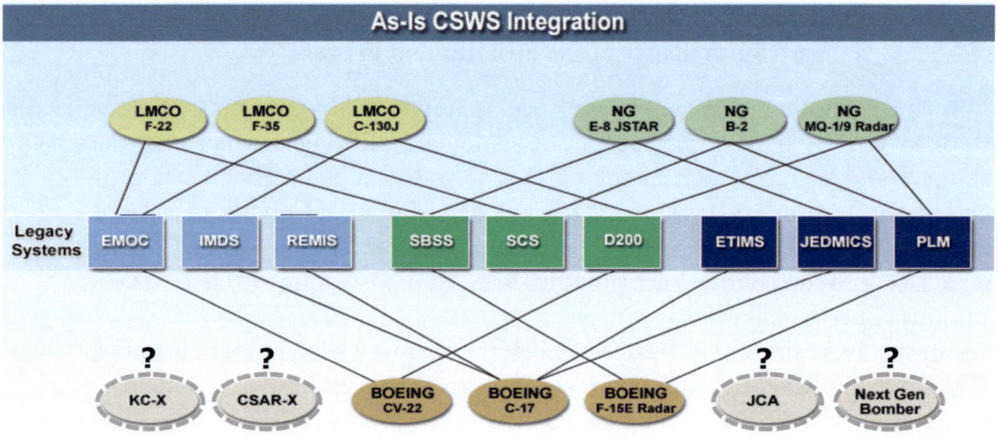

FIGURE 6-3
As-is CSWS integration. SOURCE: Grover L. Dunn, Director of Transformation, Deputy Chief of Staff for Logistics, Installations and Mission Support, Headquarters United States Air Force. "Expeditionary Logistics for the 21st Century (eLog21)." Presentation to the committee, January 17, 2011.

PROVIDING FOR CONTINUED INCORPORATION OF TECHNOLOGY FOR SUSTAINMENT

Overview

As noted in Chapter 5, to achieve ILCM, sustainment needs to be built into technology development at all stages. It is important for future systems to not only consider sustainment technologies and development processes prior to Milestone A and Milestone B as a consideration between performance and total life-cycle cost, but also to recognize that there will be opportunities for incremental insertion of technologies to address sustainment issues that arise with operation and the emergence of new capabilities.

The ALCs generally recognize and appreciate the Air Force Research Laboratory's (AFRL's) capabilities. At the same time, however, there does not appear to be a current, institutionalized process to links the ALCs to AFRL. Without addressing the full range of specific, sustainment-related technologies that may become available, the following sections discuss three general technology areas—software systems, air vehicles and engines, and integrity programs—that could lead to advances in Air Force sustainment efforts.

Software Systems

Chapter 4 includes an extensive discussion of software systems, in which potential future sustainability issues relating to the increasing resource requirements were identified. Future systems will continue to expand software-provided capability and the total delivered software base in lines of code. As a result, it is crucial that software sustainment planning be accomplished early in the weapon system development phase, and that software sustainment professionals be actively involved in the development process.

Air Vehicles and Engines

Chapter 5 includes examples of relevant sustainment technology areas for air vehicles and engines, including several areas for long-term research. It is critical that future systems take advantage of ongoing science and technology (S&T) programs for sustainment technology, as well as lessons learned from current platforms, to execute a program that matures relevant sustainment technologies in the development phase prior to Milestone B. In particular such future systems should recognize the potential for much longer extended service lives and should address up front the technologies that will improve fatigue life in structure, materials, and engine components.

Integrity Programs

New designs will benefit from Aircraft Structural Integrity Program (ASIP), Engine Structural Integrity Program (ENSIP), and Force Structural Integrity Program (FSIP) procedures to insure the structural integrity of new aircraft systems, which, in turn, minimizes sustainment cost. The current interest in balancing sustainment requirements with performance requirements provides the opportunity to include the proven technology tools from the earliest design stages and prepare for systematic approaches to sustainment.

THE UNIQUE SUSTAINMENT ASPECTS WITH RESPECT TO RAPIDLY FIELDED SYSTEMS

Sustainability issues related to systems developed under advanced development projects (ADPs) are unique. A recent Defense Science Board (DSB) report states:

> All of DoD's needs cannot be met by the same acquisition processes. Desired systems, capabilities, and material may have major variations in urgency, technology maturity, and life cycle considerations. Collectively, these will dictate the appropriate procedures needed for effective acquisition and timely delivery. To facilitate these goals, the DoD needs to codify and institutionalize "rapid" acquisition processes and practices that can be tailored to expedite delivery of capabilities that meet urgent warfighter needs.[22]

The report further recommends that:

> While there may be instances early fielding of prototypes with Contractor Logistics Support is appropriate, the risks must be well understood and parallel efforts should be in place to mature the technology and to insure that sustainment elements are adequate for the system life cycle.[23]

All ADPs require initial funding to develop the basic tenets of sustainment as these systems are developed. Once a decision is made to take an immature system into the operational environment the Air Force should develop and field a support system in parallel and with the same urgency.[24]

---

[22]DSB. 2009. Fulfillment of Urgent Operational Needs. July. Washington, D.C.: Office of the Under Secretary of Defense for Acquisition, Technology, and Logistics. pp. viii-ix. Available at http://www.acq.osd.mil/dsb/reports2000s.htm. Accessed November 22, 2010.

[23]Ibid.

[24]Briefers to the committee stressed the importance of the Air Force and other agencies being able to move technology rapidly into the theater to respond to the warfighter's critical needs, and cautioned against inhibiting the generation and deployment of vital responses. They did, however, express concern about instances when the preparation of system support was less than adequate and the catch-up by either the Air Force or the contractor caused issues during transition to normal

A primary issue with many of the rapid fielding programs currently in service stems from the full reliance on contractor proprietary designs and the lack of any level of sustainment planning even during transition to operational use. As a result the Air Force lacks data to effectively maintain and repair the aircraft outside of the contractor, and post-system acquisition procurement of these data was mentioned to be unaffordable.[25] To this end, the Air Force is establishing additional considerations for its rapid fielding programs. These include tailored logistics health assessment criteria in up-front program planning, and approval and closer interaction with the Air Force acquisition offices to ensure contractual hooks are in place for procurement of repair data if the system is ultimately fielded. Although the Air Force is addressing sustainment policy in line with the DSB report recommendations and current tailored sustainment planning processes, more actions are required to assure long-term sustainment of rapidly fielded platforms.

Finding 6-7. Emerging ADP systems that are rapidly introduced into the operational environment have not had the required sustainment support system development performed prior to deployment into a combat operational environment or training operational environment.

Recommendation 6-7. The Air Force should provide all ADPs at least a minimal level of initial funding to identify the long-term support concept and to develop the basic tenets of the support systems. Once a decision is made to take an immature system into the operational environment, the Air Force should seek funding to develop and field a support system as rapidly as possible.

COMMERCIAL AVIATION PRACTICES FOR AIR FORCE CONSIDERATION

It is particularly interesting to understand how the air transport industry approaches the issue of sustainment. The Air Force, unlike the commercial aviation industry, is required to perform variety of missions that that necessitates a wide spectrum of platforms (e.g., fighters, bombers, tankers, cargo, rotary wing, and commercial derivative special purpose aircraft). Yet, there are valuable applications from the commercial sector that can be applied to Air Force sustainment.

operations and training operations. This concern extended to both logistics support and training of support personnel. It was also noted that the form of acquisition used in many of these systems left the Air Force with extremely poor bargaining positions relative to support costs and a dependence on contractor support even though it was not the best value. Certain conditions demand rapid fielding such as those for the Predator, Reaper, and Global Hawk remotely piloted aircraft.

[25] Mark Slasor, Director of Logistics, Intelligence, Surveillance, Reconnaissance and Special Operations Forces Directorate (ASC/WI). "Perspective on Terms of Reference." Presentation to the committee, December 8, 2010.

Among these are consistent data collection, airworthiness certification processes, and maintenance/engineering standards to determine frequency and depth of maintenance actions.

The operational equivalent to sustainment in the commercial air transport industry leads to an airworthiness certification. The definition of, and final authority for airworthiness rests, by law, with the Federal Aviation Administration (FAA) as the regulatory body for aerial activity within the air spaces under U.S. jurisdiction. The FAA has established regulations that govern the certification of individuals and organizations necessary to the conduct of air commerce and other air operations, as well as regulations that govern flight rules and general operation of aerial vehicles. As witnessed following the recent Southwest Airlines crown skin failure in April 2011, the FAA regulatory process carries significant weight.

Each aircraft maintenance inspection/action standard and interval assures that aircraft are operated throughout their lifetimes in a condition and configuration consistent with current certification standards for each aircraft type. All persons, engineering, processes, parts, materials, training, certification, and servicing/cleaning fluids (i.e., sustainment practices) exercised on every aircraft must comply with standards approved by the FAA. As the committee reviewed several programs, there was no testimony to similar rigorous application of standards to general aircraft types (e.g., all fighters, all cargo).

Delegation

Although the FAA retains final authority for the determination of airworthiness, it has delegated the authority to verify that the conditions for airworthiness for each particular aircraft are met for all flight operations to the OEM and/or the airlines (Operator), Fixed Base Operator (FBO), or Maintenance, Repair and Overhaul (MRO) facility. This delegation provides a shared approach to airworthiness assurance between the FAA, OEM, and Operator, often referred to as the "3-legged stool" for safety. This triad alignment is made viable through long-established, efficient, and effective processes for communications, planning, and shared action plans to address airworthiness problems. Near the end of this study, the committee learned that the Air Force recently placed airworthiness certification for all aircraft under the Aeronautical Systems Center, Directorate of Engineering. The committee believes this is an appropriate step forward to begin to achieve some standardization of the results as well as the administrative and engineering processes that lead to the appropriate results.

The technical departments of each airline act as the delegated organization with sole responsibility, or Single Process Owner in DoD terms, for airworthiness. As such, the senior officer for maintenance operations has the responsibility, and all required authority, to assure that airworthiness objectives are met. Representatives from the airline maintenance departments generally participate in the contractual negotiations with the OEM and suppliers to assure that long-term sustainability goals are addressed through contractual provisions and guarantees. Material/supply, engineering, and maintenance finance functions generally fall under the purview of the maintenance department. Thus, technical departments have the controls for all personnel, material, and the financial resources (but not without corporate constraints) necessary to exercise its responsibility to assure airworthiness.

Aging Aircraft Sustainment

The industry process to address aging begins with the OEM at the design and certification phase for a new aircraft type. The OEM begins multi-life-cycle testing under conditions replicating the operating environment of the aircraft to determine the effects on the long-term durability of the materials and processes incorporated in the aircraft's design. These life-cycle findings are supplemented by real-life conditions of operating "fleet leader" aircraft as each type enters into operation. Airlines generally identify the fleet leaders in their own fleet, as well, and conduct additional maintenance inspections and actions as recommended by their own engineering organizations or the findings of the fleet leader program.

Maintenance Program Development

The FAA establishes and provides leadership for a team of experts, known as a Maintenance Review Board (MRB), to develop appropriate initial maintenance requirements for newly proposed aircraft. Industry Steering Groups (ISGs) are formed, under the auspices of the MRB. The methodology for the analysis and development of an initial maintenance plan is contained in guidance material derived from input by FAA, CAA/UK, AEA, U.S. and European aircraft and engine manufacturers, and U.S. and foreign airlines. The current revision of that document is MSG-3, and it provides guidance to identify maintenance and structural significant items and to review all structural elements, systems, and components of the aircraft to determine initial maintenance and inspection schedules, as well as servicing and test requirements. Recommendations are submitted to the MRB as part of the certification process for the initial maintenance program for a new or variant of an existing aircraft type.

The MSG-3 process continues throughout the life of the aircraft type and focuses on determination of hidden failures and the consequences of failure. The results of the MSG-3 review determine when aircraft structure, equipment, or components should be replaced or, on some occasions, when redesign is required to assure the required level of safety and constitute the means for revising maintenance programs over time. Consequently data collection, analysis, and action plan development for critical maintenance/operations activities are global initiatives that have proven to be uncommonly effective in maintaining exceptionally high levels of air worthiness. It is from this departure point that the Air Force efforts for a sustainment enterprise could begin. It was the general observation of the committee that standardized processes such as MSG-3 are not widely used across the Air Force. While MSG-3 or similar processes may be employed, they are used to varying degrees and with various amounts of discipline on a system by system basis.

Finding 6-8. The Air Force does not currently use a standardized data collection and engineering process for its aircraft. The commercial aviation industry approach to airworthiness has advantages that may serve the Air Force equally well.

Recommendation 6-8. The Air Force should investigate the advantages of applicable commercial policies, engineering efforts, data collection and analysis, and governance structures to manage and improve its sustainment activities as it moves toward an enterprise sustainment organization.

Recommendation 6-9. The Air Force should consider incorporating commercial-like engineering models and data collection and analysis techniques into the appropriate future platforms and contractually require that these efforts be compatible with Air Force data systems.

CONCLUDING THOUGHTS

The Air Force and the supporting contractors have strong capabilities in both technology development and maturation as noted in Chapter 5 and in specific cases in this chapter. When specifically focused, these capabilities are applicable to introducing sustainment features into future designs to complement the traditional focus on system performance. The successful application of lessons learned from field experience is exemplified in some recent designs, and evolving human factors techniques provide tools for integrating new maintenance functions and personnel capabilities. The experience base from the most recently deployed systems containing special emphasis on LO features and significantly more use of software also provide a wealth of data for incorporating sustainment capabilities into future designs. During a discussion with two contractors, the committee learned that much of the activity that resulted in incorporating sustainment features into the systems was initiated by the contractor in one case and by detailed sustainment requirements from another service in the other. This again emphasizes the importance of having strong sustainment involvement in all phases of Air Force procurements including the concept and design phases, in which the requirements and configurations are established, and the deployment and support phase, in which sustainability trends and drivers are identified.

The Air Force sustainment enterprise process is enormously complex, and there is a need for the Air Force to address change with a comprehensive and inclusive management approach. Many of the recommendations made throughout the report address specific areas of the Air Force sustainment enterprise, and these recommendations can produce a positive improvement in operational effectiveness, cost efficiency, systems availability, and overall responsiveness. A true system-of-systems approach, however, that prioritizes and balances the implementation of each of these recommendations will be required for the Air Force to achieve these goals.

Finding 6-9. The Air Force has the capability within its existing leadership structure, management acumen, and support tools to achieve success in its sustainment enterprise by moving forward with a system-of-systems approach.

Recommendation 6-10 The Air Force should utilize a systems approach in addressing the implementation of the recommendations of this report to achieve a proper balance between organizational structure, management techniques, and performance objectives.

Appendixes

Appendix A

Biographical Sketches of Committee Members

S. Michael Hudson, *Co-Chair*, is currently Chairman of I Power Energy Systems. Prior to that he was Vice Chairman, Rolls-Royce North America Holdings, a position he assumed in early 2000 and continued until his retirement in spring 2002. He also held the positions of President and Chief Executive Officer, Chief Operating Officer, and Chief Financial Officer with Rolls-Royce Allison, following its acquisition by Rolls-Royce in 1995. He served on the Boards of several joint venture companies in which Rolls-Royce Allison had interest. After he graduated from the University of Texas with a degree in mechanical engineering, Mr. Hudson was employed by Pratt & Whitney Aircraft from 1962 to 1968, working in aircraft engine design, installation, and performance, engine development and demonstration, and industrial and marine engine application engineering. Mr. Hudson is a fellow of the Society of Automotive Engineers and the Royal Aeronautical Society, an honorary fellow of the American Helicopter Society, and an associate fellow of the American Institute of Aeronautics and Astronautics. Mr. Hudson served on the American Institute of Aeronautics and Astronautics Propulsion Committee, the American Helicopter Society Propulsion Committee, the Board of Directors of the National Association of Manufacturers, and the Board of Directors of Indianapolis Water Company, and he was Chairman of the American Helicopter Society Board of Directors. Mr. Hudson was a member of the Society of Automotive Engineers (SAE), and he served as Chairman of the SAE's Aerospace Council and on its Aerospace Program Office and Finance Committees. He received the SAE Franklin W. Kolk Air Transportation Progress Award and the Royal Aeronautical Society British Gold Medal. Publications range from technical work on propulsion to defense procurement and business initiatives. Mr. Hudson served on Air Force and Department of Defense review groups, and he was a member of NASA's Aeronautics Advisory Committee and the Subcommittee on Rotorcraft Technology and chaired the Propulsion Aeronautics Research and Technology Subcommittee. He also served on several National Research Council (NRC) Committees, including the Committee on Aeronautics Research and Technology for Environmental Compatibility, the Committee on Analysis of Air Force Engine Efficiency Improvement Options for Large Non-Fighter Aircraft, and the Committee on Materials Needs and R&D Strategy for Future Military Aerospace Propulsion Systems and has been a member of the NRC's Aeronautics and Space Engineering Board.

Michael E. Zettler (Lt Gen, USAF, Ret.), *Co-Chair*, is currently the Principal of a consulting firm, Z-Zettler Consulting, and is also affiliated with the Durango Group. General Zettler (retired) served in the U.S. Air Force for more than 33 years and retired in January 2004. He was last assigned at the Pentagon in Washington, DC, as the Air Force Deputy Chief of Staff for Installations and Logistics. He was responsible for policy, resourcing, and overseeing all Air Force logistics, including all facets of weapon systems sustainment, civil engineering, services, and communications operations. He brings a wealth of experience from the private sector and the Department of Defense. He has special skills in logistics, information technology, and business operations. He recently served as a Senior Advisor to the Air Force Chief of Staff and the Air Force Deputy Chief of Staff for Installations, Mission Support and Logistics. Prior to starting his own successful company, he was a Senior Vice President of SI International Logistics' business unit and the immediate past President of LOGTEC, a private firm specializing in logistics applications and information technology management. General Zettler earned a B.A. in chemistry from the University of Cincinnati and a master's in management from Troy State University, and he graduated from the Industrial College of the Armed Forces at Ft. McNair, Washington, D.C. He has completed advanced executive management courses at Harvard University as well as at numerous professional and management development academies.

Meyer J. Benzakein, a member of the National Academy of Engineering, is currently Director of the Propulsion and Power Center at the Ohio State University (OSU). He assumed this position in July 2010. He recently completed a 5-year tenure as Chair of the Aerospace Engineering Department at OSU. He assumed this position in early 2005 after retiring from General Electric Aircraft Engines (GEAE), where he was responsible for the Research and Technology Development and New Product Creation during the past 10 years. At General Electric, he led the research effort in computational aerodynamics, aeroacoustics, aeromechanics, and combustion. He was responsible for building one of the strongest research organizations in the nation. This work helped propel GEAE to the leading position in the aero propulsion field. Dr. Benzakein is currently a Co-Director of the Ohio Center of Advanced Propulsion and Power, a university alliance dedicated to the research and development of new technologies focused on turbine-based systems. He is a fellow of the American Institute Astronautics and the Royal Aeronautical Society, and he received the Gold Medal of Honor from the Royal Aeronautical Society in 2001. He is the recipient of the 2007 AIAA Reed Aeronautics Award. He has served on many national academy, industry, and government advising panels and received an honorary doctorate from the University of Poitiers, France, in 2006.

Charles E. Browning is the Torley Chair in Composite Materials and Chair of the Chemical and Materials Engineering Department at the University of Dayton. He received his B.S. in chemistry from West Virginia University, his M.S. in chemistry from Wright State University, and his Ph.D. in materials engineering from the University of Dayton. Previous to his work at the university, he was Director of the Materials and Manufacturing Directorate of the Air Force Research Laboratory. Dr. Browning was responsible for the planning and execution of the Air Force's advanced materials, processes, and manufacturing and environmental technology programs to support all elements of Air Force acquisition and sustainment. Dr. Browning was also responsible for interfacing these specific areas throughout the corporate Air Force and Department of Defense. At the Materials and Manufacturing Directorate he headed an organization of approximately 530 government employees with a yearly budget of nearly $400 million. Dr. Browning began his career with the Air Force in 1966 and has held various senior technical and management positions within the laboratories. He was appointed to the Senior Executive Service in 1998. He has numerous awards including Outstanding Engineer and Scientist Award from the Affiliates Society Council of Dayton, the Materials Laboratory Cleary Award for Scientific Advancement, the Materials Laboratory Schwartz Award for Engineering Excellence, the Materials Directorate Management Excellence Award, and the 2002 Meritorious Executive Presidential Rank Award. He is a member of the American Chemical Society and a fellow of the Society for the Advancement of Material and Process Engineering.

Dianne Chong is Vice President of Materials Assembly, Factory & Support Technology in the Boeing Engineering, Operations & Technology organization. In this position she leads the organization responsible for development and support of manufacturing processes and program integration for the Boeing Enterprise. Prior to this she was Director of Materials & Process Technology for Boeing Commercial Airplanes. Dr. Chong was also the Director of Strategic Operations and Business for IDS Engineering. In this capacity, she was the lead director for defining and implementing a solid strategy for all Boeing engineering. She has also been the department head/team leader of, or liaison for, the materials and process and process control groups in Phantom Works and Integrated Defense Systems. Dr. Chong received bachelor's degrees in biology and psychology, master's degrees in physiology and metallurgical engineering, and a Ph.D. in metallurgical engineering from the University of Illinois. She also completed an executive master's degree of manufacturing management at Washington University. Dr. Chong served as the St. Louis representative to Military Handbook 5, where she chaired the Aerospace Users' Group and the Titanium Casting Group. Dr. Chong is a member of TMS, AIAA, ASM International, SME, SWE, Beta Gamma Sigma, and Tau Beta Pi. She has been recognized for managerial achievements and as a diversity change agent and as an outstanding alumna of University of Illinois in 2006. Dr. Chong has been a member of the National Materials Advisory Board. She served as President (2007-2008) and on the Board of Trustees and is a fellow of ASM International. Dr. Chong is currently serving on the National Research Council's Board on Global Science and Technology and is a commissioner to the ABET EAC. In 2010, she received the AAEOY award for corporate management. She was also elected to the Fellows of SME in 2011.

David E. Crow, a member of the National Academy of Engineering, is currently a Professor of mechanical engineering at the University of Connecticut and is also consulting for power generation and aircraft engine companies. He retired as Senior Vice President of Engineering from Pratt &Whitney in April of 2002. Dr. Crow was named Senior Vice President of Pratt & Whitney's engineering organization in May 1997. His responsibilities included the design, development, and in-service support of Pratt & Whitney's commercial, military, and power generation engines as well as both solid and liquid rocket engines. He also led the research and development of advanced technologies systems to meet future aerospace requirements. Dr. Crow also was Senior Vice President for Pratt & Whitney's Large Commercial Engines organization, which included the high-thrust family of products: PW4000 and JT9D propulsion systems. He was responsible for engine program management, systems design and integration, product planning, and business management. Dr. Crow jointed Pratt & Whitney in 1966 as an analytical engineer and has broad experience in multiple engineering disciplines and manufacturing. Dr. Crow is a member of the Connecticut Academy of Scientists and Engineers. Dr. Crow belongs to the American Society of Mechanical Engineers, Society of Automotive Engineers, and the American Institute of Aeronautics and Astronautics and to the honor societies Pi Tau Sigma, Phi Eta Sigma, Tau Beta Pi, Phi Kappa Phi, and Blue Key. Dr. Crow is on the Engineering Advisory Board at University of Connecticut and is member of the University of Missouri-Rolla Academy of Mechanical Engineers. He received his Ph.D. in mechanical engineering from the University of Missouri-Rolla.

Frank R. Faykes (Maj Gen, USAF, Ret.) is currently an independent consultant. He retired from the Air Force as Deputy Assistant Secretary for Budget, Office of the Assistant Secretary of the Air Force for Financial Management and Comptroller, Headquarters U.S. Air Force, Washington, DC. He was responsible for planning and directing Air Force budget formulations that annually totaled more than $136 billion. He led a staff of civilian and military financial managers who developed, integrated, and defended Air Force resource requests to the Office of the Secretary of Defense, the Office of Management and Budget, and Congress. He managed and executed funding that supported military operations and Air Force priorities at the direction of the Secretary of Defense and Congress. General Faykes was commissioned through the Air Force ROTC program at Virginia Tech and entered active duty in 1976. He has served in comptroller and command positions at all management levels—wing, major command, and Air Force headquarters. Prior to assuming his current position, General Faykes was Director of Financial Management and Comptroller, Headquarters Air Force Materiel Command, Wright-Patterson Air Force Base, Ohio. In this position, he was responsible for the oversight of more than $40 billion of the Air Force budget. General Faykes holds a M.S. in management from Troy State University and is a 2001 Seminar XXI Fellow from the Massachusetts Institute of Technology.

John T. Foreman is currently Chief Engineer for Air Force Programs at Carnegie Mellon University's Software Engineering Institute (SEI), where he is responsible for providing direct acquisition and technical support to space systems, command and control systems, and enterprise system acquisition programs in the Air Force. Prior to being named Chief Engineer, he was Director of the Dynamic Systems Program and had management, technical, and financial responsibility for the COTS-based Systems initiative, Performance Critical Systems initiative, and the TIDE (Technology Insertion, Demonstration, and Evaluation) project. During 1991-1995, Mr. Foreman completed a by-government-request assignment as the Program Manager for the Software Technology for Adaptable and Reliable Systems (STARS) program at the Defense Advanced Research Projects Agency (DARPA). Under his direction, the program defined and successfully demonstrated "product lines"—a new paradigm in software development. Prior to his assignment with DARPA, Mr. Foreman was Manager of SEI's Ada and STARS support group, working toward removing technical and managerial impediments to the adoption of Ada and also toward developing and transitioning new software engineering design approaches/paradigms facilitated by Ada. Mr. Foreman was the primary author of the *Ada Adoption Handbook: A Program Manager's Guide*. Prior to joining SEI, he was a Branch Manager for Texas Instruments (TI), responsible for planning, directing, and executing TI's Ada technology insertion strategy, including support to embedded, mission-critical development programs in avionics, command and control, and missile systems. While serving in the U.S. Air Force, Mr. Foreman's assignments focused on the development and maintenance of large mission-critical software systems. A 1973 graduate of the U.S. Air Force Academy, he holds an M.S. in computer science from the Florida Institute of Technology.

Wesley L. Harris, a member of the National Academy of Engineering, is Associate Provost, Charles Stark Draper Professor, and former Head of the Department of Aeronautics and Astronautics, and founding Director of the Lean Sustainment Initiative at Massachusetts Institute of Technology. His research focuses on theoretical and experimental unsteady aerodynamics and aeroacoustics, computational fluid dynamics, hemo-dynamics, sustainment of complex systems, and federal government policy impact on procurement of high-technology systems. Prior to this position he served as the Associate Administrator for Aeronautics at NASA. He has also served as the Vice President and Chief Administrative Officer of the University of Tennessee Space Institute and Dean of Engineering, University of Connecticut. Dr. Harris earned a Ph.D. in aerospace and mechanical sciences from Princeton University. Dr. Harris has served on numerous NRC studies and is a former member of the Air Force Studies Board.

Howard F. Hetrick has been with Northrop Grumman (NG) for 37 years and has significant supportability and managerial experience. He is currently assigned to the F-35 Program as the NG Performance-Based Logistics' Planning and Development and Sustainment Execution Integrated Product Team Lead. Mr. Hetrick's contributions to the program have played a critical role for both NG as a Team Mate and the Lockhedd Martin Aeronautics F-35 Team. Mr. Hetrick holds the position of Supportability Engineering Technical Fellow within the corporation. He served 4 years with the U.S. Air Force as an aircraft maintenance technician and supervisor. Mr. Hetrick has worked for both U.S. and international programs, developing new and innovative approaches for supporting today's weapon systems. He has played a key role in the development of the B-2 support system and the fielding of the aircraft. His professional involvement includes organizations concerned with acquisition reform initiatives, commercial standards development, commercial practices, logistics long-range planning, logistics business process re-engineering, electronic commerce and data interchange, and business enterprise development. He has been active in a variety of professional societies including the Defense Sustainment Consortium (DSC), Society of Automotive Engineers (SAE), Society of Logistics Engineers (SOLE), and National Defense Industry Association (NDIA). He holds a B.S. in Aviation Management and a M.S. in Logistics Engineering.

Clyde Kizer was President and Chief Operating Officer of Airbus North America-Customer Service from 1992 until his retirement in April 2004. At Airbus, Mr. Kizer was responsible for all aspects of technical support for Airbus North America customers, including field service, spare parts, training, engineering, quality programs, and vendor monitoring. He also coordinated on behalf of Airbus with representatives of U.S. airlines and with government officials in various activities designed to assure the safe and economical operation of the air transportation industry in North America. Prior to joining Airbus in 1992, Mr. Kizer held senior executive positions with Midway Airlines, the Air Transport Association of America, and United Airlines. He retired as a Captain from the U.S. Navy following 22 years of service as a naval officer, aviator, experimental test pilot, and combat pilot with more than 9,000 flight hours during the Viet Nam era.

Thomas A. McDermott, Jr. is the Director of Research and Deputy Director of the Georgia Tech Research Institute (GTRI), where he is the Executive Manager for GTRI's $200 million portfolio of research programs across eight research labs. He has 27 years of background and experience bridging applied research and development, major system development, project management, and executive management. He is a Principal Instructor in the Georgia Tech College of Engineering's Professional Masters Degree in Applied Systems Engineering program, and he teaches in several continuing education courses in the areas of systems engineering, electronic warfare, and project management. He has current research interests in analysis of complex systems, systems engineering methods and tools, and leadership disciplines for engineering teams. Prior to joining GTRI, Mr. McDermott developed a large breadth of experience in both technical and management disciplines at Lockheed Martin, culminating in the role as Chief Engineer and Program Manager for the F-22 Raptor Avionics Team. While at Lockheed Martin, Mr. McDermott provided technical and management leadership in avionics and computing architectures, software architecture, computer networks, fault tolerant systems, and secure systems. He has extensive knowledge and expertise in systems and software engineering and has served on a number of government independent assessment teams for major acquisition programs. Mr. McDermott holds a B.S. in physics and an M.S. in electrical engineering, both from GTRI.

Lyle H. Schwartz, a member of the National Academy of Engineering, is a Senior Research Scientist with the Department of Materials Science and Engineering at the University of Maryland. He was Professor of materials science and engineering at Northwestern University for 20 years and Director of Northwestern's Materials Research Center for 5 of those years. He then became Director of the Materials Science and Engineering Laboratory at the National Institute of Standards and Technology, where he served for more than 12 years. His experience there included metals, ceramics, polymers, magnetic materials, techniques for characterization, and standardization of these characterization techniques, and his responsibilities included management of the R&D agenda in the context of a government laboratory. Dr. Schwartz subsequently assumed responsibility for basic research on structural materials of interest to the U.S. Air Force in addition to the areas of propulsion, aeromechanics, and aerodynamics. He then completed his government service as Director of the Air Force Office of Scientific Research with responsibility for the entire basic research program of the Air Force. His current interests include government policy for R&D, particularly for materials R&D, materials science education at K-12 and university levels, and enhanced public understanding of the roles and importance of technology in society. Dr. Schwartz received his Ph.D. in materials science from Northwestern University.

Bruce M. Thompson leads the System Readiness and Sustainment Technologies Department at Sandia National Laboratories in Albuquerque, NM. His experience includes analyses to support lifecycle sustainment decisions for a wide variety of Department of Defense (DoD) legacy and current acquisition programs including the F-35 Joint Strike Fighter, the Army's Program Executive Officer (PEO) Integration, PEO Ground Combat Systems, the PM Apache Helicopter, the Missile Defense Agency's Airborne Laser, the Navy's PEO Littoral Combat Ships, and the Advanced Cruise Missile. Mr. Thompson also leads Sandia's Center for System Reliability. In addition to his DoD experience, Mr. Thompson has addressed sustainment challenges in the industrial sector, the energy sector (wind, coal, nuclear, and high-power electronics), and the Department of Energy's nuclear weapons enterprise. He has more than 25 years of technical and management experience developing and applying advanced modeling, simulation, and optimization capabilities. As a Distinguished Member of the Technical Staff at Sandia, Mr. Thompson led development of the System of Systems Analysis Toolset (SoSAT) for the U.S. Army's Future Combat Systems Program. He also led the design and development of the Support Enterprise Model (SEM), a global-scale integrated military logistics simulation toolset. Mr. Thompson has a B.Sc. in civil engineering from Loughborough University of Technology and a M.Sc. in structural mechanics from the University of Wales, Swansea.

Raymond Valeika is an independent consultant advising major companies in aviation matters. He is an internationally recognized aviation operations executive with more than 40 years of experience managing large airline maintenance operations, equally comfortable in the United States and abroad dealing with regulators, manufacturers. and employees. Mr. Valeika retired as Senior Vice President for Technical Operations for Delta Airlines, where he directed a worldwide maintenance and engineering staff of more than 10,000 professionals, maintaining a fleet of nearly 600 aircraft. Through his leadership and focus on continuous improvement of the human processes in aviation maintenance, Delta Technical Operations consistently rated at the top of the industry for performance benchmarks in the areas of safety, quality, productivity, and reliability. During his tenure, he created Delta TechOps as an entity, which has become one of the leading maintenance, repair, and overhaul (MRO) service providers in the aviation industry. Under his leadership, TechOps implemented Six Sigma and LEAN management techniques, making it one of the most productive work groups in the industry. TechOps has been honored with numerous awards, including the FAA-AMT Diamond Award, Georgia Oglethorpe Award, Aviation Week & Space Technology MRO Innovation Award, and Star status in OSHA's Voluntary Protection Program (VPP). Mr. Valeika was honored with the Air Transport Association's Nuts and Bolts award because of his leadership in the aviation industry. In addition, his leadership of the "human" side over the years was recognized by a Humanitarian Award from the Community Mayors of New York, New Jersey, and Connecticut, and a Laurel from *Aviation Week* and *Space Technology*. In October of 1999 he received the Marvin Whitlock Award from the Society of Automotive Engineers because of his accomplishments and long-term leadership within the aeronautical engineering and commercial aviation industries. Most recently the Aviation Week Group honored him with a lifetime achievement award. He has lectured at many universities including the Massachusetts Institute of Technology, Yale, and Georgia Tech and has been on various committees for the National Academies of Science. He is former member of the Aeronautics and Space Engineering Board of the National Research Council. Currently he is on the Board of Directors of Flight Safety Foundation, AerCap Inc., and SRT. Prior to working at Delta, he was the Senior Vice President of Technical Operations at Continental Airlines and Vice President of Maintenance and Engineering at Pan AM. He graduated from St. Louis University with a degree in aeronautical engineering.

Appendix B

Meetings and Participating Organizations

SITE VISIT
OCTOBER 8, 2010
THE PENTAGON
WASHINGTON, D.C.

Discussions on Expectations for Study

Lieutenant General Loren M. Reno, Deputy Chief of Staff for Logistics, Installations and Mission Support, Headquarters U.S. Air Force

Ms. Deborah K. Tune, Principal Deputy Assistant Secretary of the Air Force for Installations, Environment and Logistics, Office of the Assistant Secretary of the Air Force for Installations, Environment and Logistics

MEETING 1
OCTOBER 20-22, 2010
ONE WASHINGTON CIRCLE HOTEL
WASHINGTON, D.C.

Co-Sponsor Discussion of Study Background and Task

Dr. Steven H. Walker, Deputy Assistant Secretary of the Air Force for Science, Technology and Engineering, SAF/AQR

Air Force Sustainment for the Future
Ms. Sue Lumpkins, Principal Deputy Assistant Secretary of the Air Force for Installations, Environment and Logistics

Developing the Right Product Support Concepts for the Future
Ms. Debra K. Tune, Principal Deputy Assistant Secretary of the Air Force for Installations, Environment and Logistics, Office of the Assistant Secretary of the Air Force for Installations, Environment and Logistics

Air Combat Command A4 and A8 Sustainment Perspectives
Mr. Eugene Collins, Deputy Director of Logistics, Headquarters Air Combat Command

Air Mobility Command/A4
Mr. Timothy Thomas, Deputy Chief, Maintenance Division, HQ Air Mobility Command

Budgeting Considerations Relating to Sustainment
Mr. Blaise J. Durante, Deputy Assistant Secretary for Acquisition Integration, Office of the Assistant Secretary of the Air Force for Acquisition

Life Cycle Sustainment
Mr. Steve Gray, Director, Global Sustainment Business Design and Enterprise Integration, Lockheed Martin Corporation

MEETING 2
DECEMBER 7-9, 2010
WRIGHT-PATTERSON AIR FORCE BASE
DAYTON, OHIO

Air Force Materiel Command/CC
General Donald J. Hoffman, Commander, Air Force Materiel Command, Wright-Patterson Air Force Base

Air Force Materiel Command/A4
Major General Kathleen D. Close, Director, Logistics and Sustainment, Headquarters Air Force Materiel Command, Wright-Patterson Air Force Base

Global Logistics with a Focus on the Warfighter
Major General Gary McCoy, Commander, Air Force Global Logistic Support Center, Scott Air Force Base

Aeronautical Systems Center/WW (Fighter/Bomber PEO)
Brigadier General Arnold W. Bunch, Jr., Director and Program Executive Office for the Fighters and Bombers Directorate, Aeronautical Systems Center, Wright-Patterson Air Force Base

Air Force Materiel Command/EN
Colonel Kurt Hall, Deputy Director, Engineering and Technical Management Directorate, Headquarters Air Force Materiel Command, Wright-Patterson Air Force Base

Aeronautical Systems Center
Colonel Arthur Huber, Vice Commander, Aeronautical Systems Center, Wright-Patterson Air Force Base

Aeronautical Systems Center/WLM (C-17)
Colonel Mark Mol, Chief, C-17 Division, Mobility Directorate, Aeronautical Systems Center, Wright-Patterson Air Force Base

Air Force Research Laboratory Sustainment Investment
Dr. Katherine Stevens, Director, Materials and Manufacturing Directorate, Air Force Research Laboratory, Wright-Patterson Air Force Base

Greybeard Assessment of the Sustainment Technology Transition Process
Dr. Vince Russo, Executive Director (retired), Aeronautical Systems Center, Air Force Materiel Command, Wright-Patterson Air Force Base

Aeronautical Systems Center/WNWPA (Propulsion)
Ms. Valerie Dahlem, Chief, Fighter Engine Programs Branch, Wright-Patterson Air Force Base

Fleet Viability Board
Mr. Francis Crowley, Director, Air Force Fleet Viability Board

Aeronautical Systems Center/WIL (Intelligence, Surveillance, and Reconnaissance and Special Operations Forces PEO)
Mr. Mark Slasor, Director of Logistics, Intelligence, Surveillance, and Reconnaissance (ISR) and Special Operations Forces (SOF) Directorate, Aeronautical Systems Center, Air Force Materiel Command, Wright-Patterson Air Force Base

Air Force Materiel Command/FMB
Ms. Kimberly Keck, Chief, Centralized Program Budget Division, Wright-Patterson Air Force Base

SITE VISIT
JANUARY 6-7, 2011
WARNER-ROBINS AIR LOGISTICS CENTER
WARNER ROBINS, GEORGIA

Overview
Major General Robert H. McMahon Commander, Warner-Robins Air Logistics Center, Robins Air Force Base, Georgia

402d Maintenance Wing (MXW) Overview
Brigadier General Lee K. Levy, II, Commander, 402 MXW

Aerospace Sustainment Directorate Overview
Colonel Chris Davis, Director, Aerospace Sustainment Directorate
C-5 Integrated Program Review
Maintenance Steering Group 3
Colonel Michael Gregg, Chief, Galaxy Division

WR-ALC Challenges and Opportunities
Ms. Kim Lynn, Director, Plans and Programs

Global Logistics–Warfighter Focus
Mr. Donald J. Bagley, Vice Director, 638th Supply Chain Management Group (SCMG)

The Defense Logistics Agency (DLA) at Robins Air Force Base
Mr. Joe Alexander, Chief of Supply, DLA Aviation

SITE VISIT
JANUARY 11-12, 2011
OKLAHOMA CITY AIR LOGISTICS CENTER

OC-ALC Strategic Goals
Major General P. David Gillett Jr., Commander, Oklahoma City Air Logistics Center, Tinker Air Force Base, Oklahoma

76 MXW Production Machine
Major General Bruce A. Litchfield, Commander, 76th Maintenance Wing

Weapon System Life Cycle Management
Colonel Mark Beierle, Director, Aerospace Sustainment Directorate

Preparing for an Uncertain Future
Colonel Lawrence Gatti, OC-ALC Plans and Programs

Facilities Infrastructure
Col Robert LaBrutta, Commander, 72nd Air Base Wing

KC-10 Contract Logistics Support
Col Mike Schmidt

Defense Logistics Agency (DLA)
Col Richard Schwing

KC-135 Programmed Depot Maintenance (PDM)
Col Robert Torick
Ms. Janis Wood

B-2 Weapon System Support Center, System Integration Lab (WSSC SIL) Brief/Tour
Lt Col Scott Bell

Tinker Today
Ms. Caysie Mercer

Diminishing Manufacturing Sources and Material Shortages and Obsolescence Challenge
Ms. Lydia Cervantes, 429 SCMS

Engine Leading Health Indicator
Mr. Ralph Garcia

To KC-135 PDM Line
Ms. Janis Wood

B-52 Weapon System Integrity Program
Mr. Jerold Smith
Mr. Mike Hostetter

E-3 Block 40/45 Brief/Tour
Mr. Bill Cain
Mr. Thomas Ramsey

B-1 High Velocity Maintenance
Mr. Steve Walker

Repair Network Integration (RNI) F101 Brief/Tour
Mr. Brenden Shaw
Mr. Brian Babin

Engine Test Cell Brief/Tour
Mr. Floyd Craft

KC-135 Flight Controls Brief/Tour
Mr. Mike Barrett

MEETING 3
JANUARY 18-20, 2011
ARNOLD AND MABEL BECKMAN CENTER
IRVINE, CALIFORNIA

NAVAIR
CAPT Mike Kelly, United States Navy, Force Material, Maintenance and Readiness, COMNAVAIRFOR N42

Expeditionary Logistics for the 21st Century (eLog21)
Mr. Grover L. Dunn, Director of Transformation, Deputy Chief of Staff for Logistics, Installations and Mission Support, Headquarters United States Air Force

Enterprise Behavior: Fundamental Changes in the Government Business Model
Vice Admiral Walter Massenburg (United States Navy, retired), Senior Director, Mission Assurance Business Execution, Raytheon Company, Retired Commander, Naval Air Systems Command

F-35 Program Office
Major General C.D. Moore, Deputy Program Executive Officer
Mr. Todd Mellon, Senior Executive Service, F-35 Director, Logistics and Sustainment

U.S. Air Force-Boeing C-17 Sustainment Partnership
Mr. Mark Angelo, Director, C-17 GSP Operations & Site Lead, The Boeing Company
Mr. Robert Tomilowitz, Executive Director, Supply Chain Management and Support Equipment, The Boeing Company
Mr. Richard (Skip) Whittington, Senior Manager, C-17 GSP Business Development, Boeing Defense, Space & Security

American Airlines
Mr. David Campbell, Vice President for Base Maintenance

Unmanned Air Systems Sustainment: Is it Different?
Vice Admiral James Zortman (United States Navy, retired), Sector Vice President Life Cycle Logistics Support, Northrop Grumman Corporation

SITE VISIT
JANUARY 31-FEBRUARY 1, 2011
OGDEN AIR LOGISTICS CENTER

Core Competencies & Priorities
Major General Andrew E. Busch, Commander, Ogden Air Logistics Center, Hill Air Force Base, Utah

Supply Discussions w/DLA & SCMG
Colonel Jim Fisher, Mr. Rick Fuit

CMZG, SCMG & DLA Discussions
Col Hall, Mr. Lengyel

Ogden Scorecard Overview
Col Kinkade, Mr. Lengyel

Commodities DSCM
Mr. Thomas Huber

Production Perspective
Ms. Jeannette Moklofsky

DLA Shop Service Center
Col Kinkade

Joint Program Management & Depot Maintenance Discussions
Brig Gen(s) Scott Staff, Director, ASD, Col Steve LaVoye, 309 AMXG/CC

MEETING 4
TEC EDGE INNOVATION CENTER
FEBRUARY 7-8, 2011
DAYTON, OHIO

Proposed AFMC ATC Process, Sustainment Technology Process (Revised)
Mr. Dan Brewer, Air Force Research Laboratory, Air Force Materiel Command, Wright-Patterson Air Force Base

Propulsion Safety and Affordable Readiness (P-SAR)
Mr. Chuck Cross, Air Force Research Laboratory, Air Force Materiel Command, Wright-Patterson Air Force Base

Aircraft Structural Integrity Program (ASIP)
Ms. Pam Kobryn, Air Force Research Laboratory, Air Force Materiel Command, Wright-Patterson Air Force Base

Co-Locates, System Support, ALC Offices
Mr. Ed Hermes and Mr. Larry Perkins, Air Force Research Laboratory, Air Force Materiel Command, Wright-Patterson Air Force Base

Greybeard Study Discussion
Dr. Vince Russo, Executive Director (retired), Aeronautical Systems Center, Air Force Materiel Command, Wright-Patterson Air Force Base

General Electric Sustainment Initiatives
Dr. Robert Schafrik

Pratt & Whitney Sustainment Initiatives
Mr. Jeff Zotti

SITE VISIT
FEBRUARY 14, 2011
THE PENTAGON
WASHINGTON, D.C.

Air Force Policies and Regulations Applicable to Sustainment
Major General Judith A. Fedder, Director of Logistics, Deputy Chief of Staff for Logistics, Installations and Mission Support, Headquarters U.S. Air Force

MEETING 5
FEBRUARY 15-17, 2011
THE KECK CENTER OF THE NATIONAL ACADEMIES
WASHINGTON, D.C.

Air Force Fleet Viability Board
Mr. Fran Crowley, Director

Office of Naval Research Processes for Technology, Development, and Insertion for Sustainment
Dr. Walter Jones, Executive Director

Weapon System Acquisition Reform Product Support Assessment
Mr. Mark Gajda, Strategies and Plans Branch Chief, Office of the Deputy Assistant Secretary of Defense (Materiel Readiness)

Congressional Panel Discussion
Mr. Peter Levine, Senate Armed Services Committee, General Counsel, Readiness & Management Support Subcommittee, Majority Lead
Ms. Lynn Williams, House Armed Services Committee, Readiness Subcommittee, Majority Lead
Ms. Vickie Plunkett, House Armed Services Committee, Readiness Subcommittee, Minority Lead

Results of Recent RAND Corporation Studies Relevant to Sustainment
Ms. Natalie Crawford, Senior Fellow
Dr. Raymond Pyles, Senior Management Scientist

Defense Logistics Agency
Ms. Kathy Cutler, Deputy Commander, DLA Aviation

SITE VISIT
MARCH 22, 2011
THE PENTAGON
WASHINGTON, D.C.

Major General Alfred K. Flowers, Deputy Assistant Secretary for Budget, Office of the Assistant Secretary of the Air Force for Financial Management and Comptroller

MEETING 6
MARCH 29-APRIL 1, 2011
FLEET READNINESS CENTER SOUTHWEST/
ARNOLD AND MABEL BECKMAN CENTER
IRVINE, CALIFORNIA

Command Overview
Captain Fred Melnick, Commanding Officer, Fleet Readiness Center Southwest

Manufacturing Program Brief
Mr. Joe Caoile, Manufacturing Program Manager

Components Program Brief
Mr. Joe Caoile, Manufacturing Program Manager

AMS/RE Lab Program Brief
Mr. Chris Root, Advanced Aircraft Technical IPT Lead

E-2/C-2 Program Brief
Mr. Joe Garcia, E2/C2 Program Manager
Mr. Elijah Scott, LCDR, Deputy Program Manager
Mr. Eric Holsti, E2/C2 Deputy Program Manager

H-60/H-53 Program Brief
Mr. Dave Kelly, Vertical Lift Program Manager

Center Barrel Program Brief
Mr. Kevin Okerman, F/A-18 Program Manager
Mr. Walt Loftus, F/A-18 Center Barrel Deputy Program Manager

F/A-18 Program Brief
Mr. Kevin Okerman - F/A-18 Program Manager

MEETING 7
MAY 10-12, 2011
J. ERIK JONSSON CENTER
WOODS HOLE, MASSACHUSETTS

Writing meeting.

Appendix C

Navy Enterprise Transformation: Working for the Greater Good

NOTE: The document in this appendix is reprinted from USN (United States Navy) and Babson Executive Education. 2007. *Navy Enterprise Transformation: Working for the Greater Good.* May. Available at http://www.thomasgroup.com/getdoc/7c79c3c9-8603-4908-ad89-0ebb0e10a67f/Navy-Enterprise-Transformation.aspx. Last accessed on August 22, 2011.

May 2007

Navy Enterprise Transformation: Working for the Greater Good

In early 2007, Admiral John Nathman, CFFC and leader of the Fleet Readiness Enterprise (FRE), considered the progress that the Navy had made over the past several years improving readiness at reduced costs through the adoption of "enterprise" behavior. The enterprise approach empowered stakeholders across multiple commands to take a holistic view of objectives and processes and act cohesively to achieve required output with greater efficiency. It countered the numerous tendencies that encouraged consumption-oriented behaviors and that subverted people's view of and ability to work toward the "greater good" of the Navy. The enterprise model involved cross-organization collaboration and decision-making, but it did not impinge on the sanctity of the chain of the command structure.

The enterprise approach emerged in the late 1990s in response to the declining state of Navy readiness, particularly in the Navy's Air Forces. Propelled by CNO Vern Clark and led by Nathman, naval aviation first focused on improving readiness in the inter-deployment training cycle and then expanded to include deployed air units and air craft carriers. As the enterprise concept took hold, the term ***Naval Aviation Enterprise*** (NAE) was officially embraced in 2004. In 2005, the surface and sub-surface forces, both of which had been engaged in embryonic enterprise activities for many years, launched their own formal initiatives: Surface Warfare Enterprise (SWE) and UnderSea Enterprise (USE). These were followed in 2006 by the formation of two additional warfare enterprises: Naval Network Warfare/FORCEnet Enterprise (NNFE) and Navy Expeditionary Combat Command (NECC). In 2006, the Navy extended the enterprise framework into the fleet and corporate management tiers, creating the Fleet Readiness Enterprise (FRE) and the Navy Enterprise, and in early 2007 it added the Navy Provider Enterprise.

Admiral Nathman, who had been involved in the development of the enterprise concept since the late 1990s, was encouraged by the significant gains in readiness and the cost savings which the

*****NB: This case study is based on interviews with participants in the various Navy Enterprises and on published documents. It is intended as an educational tool to facilitate discussion and learning. Although every effort has been made to ensure the accuracy of the information, the case does not purport to be a definitive historical record. Any factual errors or misrepresentations of individuals' actions or words are solely the responsibility of the author.***

Sam Perkins, Babson Executive Education, prepared this case under the direction of VADM Phil Quast and RADM Harry Quast.

BABSON
EXECUTIVE EDUCATION

initiatives were starting to produce. Yet, he was also concerned about the many potential forces that could derail progress or thwart success.

> *The five warfare enterprises are at different stages of maturity and even the most mature – naval aviation – is not even halfway there in terms of what it can accomplish. The rest have a ways to go.*

Consumption Culture and Strategic Misalignment

The enterprise activities that emerged in the late 1990s and early 2000s were in large measure a response to conditions fostered by a culture of (inefficient) consumption and a lack of alignment between requirements and processes – a deeply-embedded way of life that influenced decisions and behavior across a wide range of activities. The consumption culture was endemic: as one CNO reportedly remarked: "*the only thing the Navy is good at is spending money.*" It could be seen as a logical response to a combination of interwoven factors: organization structure, tribal culture and customs, financial/budgetary systems, and reward metrics. Enterprise activities needed to challenge and overcome one or more of these factors to achieve success.

- **Structure and Tribal Culture:** The command structure fostered a stovepipe organizational mindset that restricted officers' perspectives to their own domains and obstructed a holistic view of processes that spanned multiple commands. This mentality encouraged people to optimize the performance of their own stovepipes without regard to the performance of the whole. Tribal allegiance, whether defined by the major tribes (surface, submarine, air) or sub-groupings, superimposed an additional constraint that limited the inclination of leaders to adopt a broad perspective.

- **Budget systems:** Budget policy, practices and systems contributed to the general consumption mindset and to inefficient, stove-piped allocation of money unaligned with a strategic understanding of requirements. The need to spend every dollar of a budget out of fear that failure to do so would result in reductions in future budgets served as a fundamental impediment to efficiency: as numerous officers stated: "my job was to spend every penny in my account."

- **Performance Measurement Systems:** Performance measurement systems were not designed to promote cost effectiveness; instead, they tended to encourage consumption and reinforce stovepipe attitudes, rather than link execution to strategic requirements. Metrics were often based on "more is better;" no one was rewarded for figuring out how to perform a task for less money. The coveted Battle E award was strongly based on consumption: the more one consumed the better the chance of winning. One of the most costly examples of misaligned performance measurement was in the Flying Hour Program. For naval aviators, the number of traps and flying hours was a critical element of a fitness report and a key metric of how they were measured against their peers. As one leader described: "*each airwing wanted to get more hours and traps than the previous group.*" Between 1990 and 1999 the number of traps and flying hour increase significantly, yet there was no defined understanding of the relationship between these metrics and their impact on squadron readiness or cost.

- **Lack of Strategic and Business Alignment:** Inefficient and ineffective consumption of resources was exacerbated by the stovepipe mentality and by the system for determining budget priorities and allocating moneys. In the 1990s, as DoD budget cuts and inflation significantly reduced purchasing power, Navy commands faced increasingly large gaps between desired and actual funding levels (i.e., perceived requirements and budgets). Eager to pursue their individual priorities, commands were reluctant to voluntarily cut programs. Resource sponsors typically submitted unbalanced budgets whose requirements exceeded funding availability in order to secure as much money as possible, regardless of how effectively it was spent. In the absence of rational planning, comptrollers made across-the-board percentage cuts based on dollars, not on a strategic understanding of how the money was being spent to generate the required output. In the words of one admiral:

It was a peanut butter approach to budgeting — spreading cuts evenly over all programs. The philosophy was "dollars at all cost; production at no consequence." Cuts were being applied to meet dollar constraints without a full understanding of the impact or an informed discussion of priorities. You would get very arbitrary and capricious decisions and be told to live with it. Everybody was trying to protect their funds and as a result none were being properly resourced and we all were failing together. It was a death spiral.

Impact of Consumption Culture

The impact of the consumption culture and its inherent inefficiencies was masked during the massive defense build-up in the Reagan years. As one Admiral remarked, "In the 80s, we had lots of money and we spent lots of money." The fall of the Berlin Wall and end of the bipolar world triggered major changes in US budget priorities and in DoD strategy, with significant ramifications for each service, which faced the need to prepare for a variety of potential scenarios.[1] The Navy's strategy for the 1990s was defined in *From the Sea*, a description of how to reengineer the Navy to deal with a more complex and uncertain world with new global political and military dynamics. Concomitant with the need to reengineer and recapitalize its forces, the Navy faced rapidly declining purchasing power. Reductions in the growth of DoD budgets, driven by the notion of the "peace dividend," combined with rising prices, produced fewer real dollars – essentially negative growth.

Budget decreases and inflationary pressures, combined with the demands of the 1990 war in Iraq, were exacerbated by the consumption-oriented behavior, creating an environment of difficult trade-offs between recapitalization and current readiness. Much of the naval aviation acquisition budget was diverted to pay for Desert Shield and Desert Storm, leading to deferred recapitalization in '92 and '93. In the mid and late 1990s, the Navy focused attention on recapitalizing naval air forces, reducing funds available for logistics accounts and impacting current readiness. Although recapitalization improved during this period, the Navy was unable to "get to the bow wave" of the effort. Consumption-oriented behavior, spending still driven by cold-war infrastructure requirements, and misguided budget priorities driven by comptroller-made decisions, all combined to sub-optimize both the acquisition process and current readiness.

Documenting the State of Navy Readiness – AMSR

By the late 1990s declining readiness became a major issue across the entire Department of Defense, as the Joint Chiefs stated in September 1998, "Our readiness is fraying and... the long term health of the total force is in jeopardy." In the Navy, attention focused on aviation, and starting in the mid-1990s, several studies documented the state and causes of naval aviation readiness. The first study was a 1995 report to the CNO, conducted by the Thomas Group, a process improvement consultancy, which was entitled the Health of Naval Aviation (HONA). The second was a study called the Aviation Maintenance-Supply Readiness Review (AMSR) that was directed by then CINCPACFLT ADM Archie Clemins under the leadership of RADM Justin (Dan) McCarthy. The AMSR found convincing evidence of declining naval aviation readiness at numerous locations: tin-foiled S3s at North Island; 70% increase in cannibalizations at CAG-14; 47% full mission capability rate for the embarked Marine Air Element aboard BELLEAU WOOD (LHA).

[1] The Global War games books in 1990 described three potential scenarios: 1. Russia retrenches and the world returns goes back to a bipolar orientation; 2) nationalism and ethnic/religious chaos; 3.) anarchy.

Many people in Navy leadership considered the issue to be simply a "parts problem," (i.e. not enough parts so planes can not get repaired), but at a March 1998 PACFLT commanders conference, ADM Archie Clemins introduced a fundamentally different perspective when he asked: "*What do we need to do to fly the hours for less dollars?*" According to McCarthy, the combination of these signals, coupled with CNO Jay Johnson's concern about F18 readiness at NAS Lamoore, persuaded Clemins to stand up a task force to investigate the readiness issue. In April 1998, with CNO-approved charter in hand, McCarthy, then Deputy Chief of Staff for Supply, Ordnance and Logistics, convened the AMSR working group, composed of a cross-section of leaders from the maintenance, supply and operational communities, to explore trends and possible causes. Clemins was a strong and visible supporter of the effort, participating in monthly VTCs and annual off-sites. AMSR operated for three-years until August 2001 and identified nineteen areas of concern. Sub-groups developed recommendations and experimented with solutions but were unable to make much headway due to resource constraints and resistance to change deeply-embedded behaviors. VADM Wally Massenburg, who joined the AMSR team in its second year, characterized the difficulty:

> *We documented the problem but all we could do was talk and tinker at the margins. There was no vehicle for dealing with of the problem because everybody was stuck in their stovepipes and in their tribes and all the dollars were being managed in financial stovepipes disconnected from one another.*

In spite of the lack of substantive progress, however, the leaders of AMSR viewed the process as a breakthrough, as McCarthy described:

> *It was an awakening to get all these people together at the table. People who don't normally get together--looking at the problems together – to begin to scope the integrated logistics and operational implications of providing fleet readiness. Those meetings fostered a dialogue that hadn't existed before. It had been all separate commands – just doing their own thing. AMSR was the foundation. The whole dialogue changed as a result of this constant drum beat of engagement focused on achieving readiness within available resources.*

The AMSR group submitted its findings to the Navy Investigator General's office, which published a report in April 2000 entitled Aviation Spares and Readiness. The IG report, coupled with other documents, observations and anecdotal evidence provided a convincing picture of the troubles confronting the Navy. A GAO Report in July 2001, which focused on two aircraft types, described some of the far-reaching impacts of reduced readiness:

> *The shortage of spare parts for the two aircraft systems reviewed not only have affected readiness but also have created inefficiencies in maintenance processes and procedures and have adversely affect the retention of military personnel. Also the maintenance practice used to mitigate part shortages masks the true impact of shortages and results in increased work for maintenance personnel, causing morale problems and dissatisfaction with military life.*

The ramifications of the Navy's readiness crisis struck home in late 2001 following the September 11 terrorist attacks on the Pentagon and World Trade Center. With the launch of Operation Enduring Freedom (OEF) in October 2001, eight carriers were ordered to deploy but there were supplies for only four and a half.

Process Improvement Initiatives – NAPPI

Responding to emergent readiness concerns, in the late 1990s several naval air commands launched initiatives which became foundational elements of the enterprise construct. One significant issue, which had been identified in the HONA report, was naval pilot training. The length of time required to move a pilot through the training pipeline had grown from the required 36 months (depending on aircraft type) to 48 months and longer, as pools of pilots accumulated at bottlenecks in the process. The training backlog and resultant shortage of new pilots resulted in extended at-sea deployments, which translated into reduced aviator retention. Different commands under NETC were responsible for different training functions[2], but no single command "owned" or had a holistic view of the entire process. In the words of one observer, "it was completely de-synchronous and leaderless." The lack of a single process owner, coupled with questions about how training funds were being spent, led the resource sponsor (N88), RADM Dennis McGinn at the time, to become involved in trying to find a solution. In contrast to conventional wisdom that "we just need to put more people in the pipeline and work harder," McGinn believed that the source of the problem lay in a lack of understanding of the process cycle. With the "Air Boss"[3] unable to take the lead, McGinn engaged the Thomas Group to help improve the process.

In early 1998, shortly after the Thomas Group was hired, RADM John Nathman relieved McGinn and assumed oversight of the effort – dubbed Naval Aviation Production Process Improvement (NAPPI) program. According to Nathman and others, the Thomas Group's strength was its ability to sustain a course of discovery that enabled leaders to understand their processes intimately and to identify the sources of problems. For the first time aviation leaders saw pilot training as a holistic system that started with fleet aviator requirements, based on rotation inside squadrons, and flowed all the way back up to the pipe to the recruitment of prospects. It became clear that the three commands involved in pilot training were making decisions that optimized their individual functions but that cumulatively subverted the entire process, creating pools of pilots between training phases (where skills atrophied as they waited) and adding months to the system. Ultimately, the initial key to producing pilots more efficiently was to induct fewer trainees into the system, thereby preventing the build-up of the pools. Both the appearance and substance of that solution, however, threatened to disrupt established equities, but Nathman, who saw the waste of the existing disconnected process, garnered support to push it through. Over several years NAPPI helped reduce the Time-To-Train (TTT) by 40 percent and increase the number of flight-ready aviators moving through the training program by 30 percent. It foreshadowed the emergence of the Enterprise concept in its use of cross-functional teams and quality-related metrics and tools, and especially, according to Nathman, in its reliance on decision-making by a group of leaders acting across chain of command lanes to pursue a common objective. Nathman described the importance of NAPPI:

> *NAPPI is an important story because it explains how you lead an Enterprise – who you defer to for leadership. We were going through the cycle time reduction/process improvement activity where there was no real leadership except the power of personality. No one – OPNAV, the Air Boss, the training commands – had the power alone to force this, yet, because everyone was invested in finding a solution, we were able to make some tough decisions. Enterprise behavior is a behavioral model – not a chain of command model. That is why people have to know where they belong. If the leadership got exposed to the right facts they could make good decisions and were willing to accept something sub-optimized in their own domain to make the larger system work better. That was the lesson we kept learning over and over again and people started seeing the results, and once we started seeing that then guys started to get it.*

At the same time that NAPPI was gearing up, aviation depots independently began to introduce business process improvement tools and activities in some of their operations. Cherry Point brought in the Goldratt Institute to apply its particular tool, Theory of Constraints, to the H46

---

[2] NETC - Naval Education and Training Command. API (Aviation Pre-flight Indoctrination) was responsible for recruitment and orientation; CNATRA (Chief of Naval Air Training) handled the bulk of the training before handing the pilots off to the RAGs (Replacement Air Group)

[3] The "Air Boss" was the senior operational aviator, a role that typically shifted between AIRLANT and AIRPAC depending upon seniority. The Air Boss was more a titular than a substantive position as it lacked the authority or power to take action across naval air forces.

helicopter line. (*See Appendix A for a description of Theory of Constraints and other process improvement tools: Lean, Six Sigma, Kaizen, PVM*). Jacksonville secured the services of GE "lean manufacturing" experts to revamp operations on the F404 engine line, and North Island implemented a hybrid practice, termed RIFLe – Relevant Information for Leadership. These initial efforts produced some gains in reduced cycle time and enhanced productivity, but for several years they remained localized and disconnected, lacking coordination and the vehicles and incentives to be expanded and replicated. Additionally, they were not linked to requirements determined by the fleet, and thus they did not necessarily help to improve overall objectives.

Confluence of Enterprise-oriented Leaders and Ideas

In 2000 Admiral Vern Clark became the first CNO with a strong orientation toward the "business" side of the Navy to match his reputation as a war fighter. One admiral stated, "Clark's key contribution was to get the whole organization thinking about the business of the Navy, and he focused huge amounts of attention there." According to another leader, Clark came on board with the conviction that improving how the Navy spent money to achieve current readiness would free up dollars for recapitalization. To that end Clark conceptualized the *Sea Enterprise* construct as a key element of SEAPOWER 21 that would generate resources through improved business practices.

To augment evidence from AMSR and other reports, Clark held a series of five CNO executive board meetings (CEBs) to discuss naval aviation readiness, with particular attention on the Flying Hour Program (FHP).[4] The cost of FHP had grown dramatically during the late 1990s, driven by a 14-19 percent per year increase in AVDLRs, and funds were often taken from acquisition accounts to bolster the FHP. The FHP was managed by the fleet (AIRPAC and AIRLANT), but squadrons were directly accountable only for POL: several squadron leaders cited their belief that the "cost" of flying was just the cost of fuel. Leaders offered different reasons for the cost escalation, from aging aircraft and obsolescence to lack of understanding of requirements. According to one participant, in CEB 3 CNO Clark became frustrated by the subject of traps and issued an ultimatum:

> *They put up the traps versus flying hours charts and Clark said: "It's your damn EGO that is driving this issue." He pointed to Nathman and said: "You sir are now in charge of this mess. Fix it or I'll fire you and get the next guy," and he walked out of the room. That was an important and liberating direction because now we had the opportunity to have a single person be the owner of the process.*

Tasking VADM Nathman, who had moved from N88 to Commander Naval Air Forces Pacific (CNAP) in August 2000, to oversee all of Naval Aviation as Commander Naval Air Forces (CNAF) created a critical underpinning of the enterprise effort.[5] It established Nathman as the lead Type Commander (TYCOM) and single process owner. Nathman was fully steeped in readiness issues not only from his NAPPI experience but also from participation in the monthly AMSR VTCs. McCarthy described the importance of Nathman's role. "Nathman could make the connection between the resource side and the operational community. Because of that continuity of leadership, he was the glue that pulled these components together."

As Nathman stood up AIRFOR, he was intent on imbuing it with more weight than the airboss construct. Nathman recruited then RADM Mike Malone, AIRLANT, to support the notion of a true,

[4] The Flying Hour Program was an item in O&M budget ($3.2 billion in FY 2000) for training flight operations and aircraft maintenance. The FHP consisted of three categories of cost: Petroleum, Oil and Lubricants (POL- 17% of total cost), Aviation Fleet Maintenance (Intermediate Level) (AFM - 29%) and Aviation Depot Level Repairables (AVDLRs -54%). AFM included consumables used in operations or smaller repairs. AVDLRs consisted of assemblies that were returned to the depot for rework.

[5] The concept of a single commander of naval air forces was part of an effort to move away from the East Coast/West Coast dichotomy and included the formation of Fleet Forces Command. Nathman was designated Commander, Naval Air Forces in October 2001.

single air force leader who was empowered to make global decisions, and the two leaders worked to integrate elements of the air forces and give it a single voice. Nathman clearly saw the endemic problems and recognized that a systemic approach was needed, as he recalled:

> *I found myself saying, "we need to have a NAPPI-like process and put it in aviation based on what we saw and based on the AMSR." And I used the AIRFOR model to start NAVRIIP, the first real enterprise approach to readiness.*

NAVRIIP

With $8 million in FHP execution year money, Nathman hired the Thomas Group to help develop and implement NAVRIIP – the Naval Aviation Readiness Integrated Improvement Program. NAVRIIP initially focused on the Inter-deployment Training Cycles (IDTC) – the 18 month period between deployments when readiness declined precipitously[6]. The program, launched in late 2001, involved flag officers from seventeen commands who participated on three cross-functional teams (CFTs) with the initial mission of improving maintenance and repair processes – implementing many of the findings of AMSR.

The NAVRIIP team included then RADM Massenburg, NAVAIR Logistics, who was a passionate proponent of cross-functional activity and helped to develop two principles that proved critical to NAVRIIP's success and that would become core building blocks of the enterprise framework. One principle focused on the use of the type model series (T/M/S) as the basis for managing process improvements, based on the recognition that types of aircraft were maintained at multiple locations and that optimizing maintenance processes required a holistic approach across those sites. The second principle was the concept of aligning processes around a core objective – a single readiness metric – that was of critical importance to the end user – the fleet. Massenburg and other NAVRIIP flags recognized that the metrics historically used to track readiness – FMC and MC (fully mission critical and mission critical) - were inadequate because they focused on near-term solutions (i.e., fixing existing casualties) and provided no leading indicator through which to improve readiness. The important indicator was determined to be aircraft ready to fly sorties, and the team designed NAVRIIP around the key single metric – *Aircraft Ready for Training*.

The three cross-functional teams coordinated NAVRIIPs efforts and allowed all aviation stakeholders to align around the single metric. CFT1 defined appropriate levels of readiness throughout the IDTC and developed training and readiness matrices tailored for each TMS. Essentially it developed "entitlements" based on fleet driven requirements, creating a "demand pull" signal to which maintenance organizations and the supply chain could respond. CFT2 was responsible for providing parts, aircraft and support equipment to squadrons through NAVSUP, DLA, BUPERS and NAVAIR to enable aviators to meet training milestones. CFT3 focused on planning and programming to ensure funding requirements were met.

The Thomas Group led NAVRIIP's efforts to map maintenance and repair processes, and to identify opportunities for improvement and for removing barriers. According to one NAVRIIP flag, the Thomas Group contributed two critical elements: its ***Process Value Management™ (PVM™)*** approach and its imposition of rigid discipline. ***PVM*** required the identification and purpose for every step in a process, the output of the process, and the development of metrics to determine progress toward that objective. After process improvements were developed, TG ***Resultants®*** forced officers to focus on the changes needed to secure the desired output, as the flag described: "They nag the hell out of us to do what we said we were going to do – they impose discipline." A key element in the mapping and barrier identification process was direct

[6] Newly deployed forces, having just completed their own under-resourced interdeployment cycle, tended to consume resources in greater quantities than required to maintain readiness, leading both to overspending in the Flying Hour Program and to excessively reduced levels of readiness in IDTC. The scenario was often characterized by a bathtub readiness chart, with low levels of readiness for most of the IDTC being the bottom of the tub.

communication between the senior officers and maintenance and supply organizations through a program called Boots on the Ground (BOG). The first BOG took place in August 2002 when Massenburg and RADM Mike Finley, NAVICP, along with other leaders, visited Naval Air Station Lemoore and spent time talking to junior enlisted service members about maintenance issues and barriers. BOG allowed leadership to see first-hand systemic problems in maintenance and supply processes, enabling feedback from front-line personnel which informed recommendations and actions taken corporately for naval aviation. BOG was also an important ingredient in fostering a tighter relationship between the end-users (the fleet) and the maintenance and supply providers.

The Enterprise Construct – A Behavioral Model

During the early stages of NAVRIIP, Massenburg drew a triangle for Nathman that depicted his concept of the relationship between the three types of stakeholders involved in the effort (*Figure 1*). The genesis of the triangle was Massenburg's experience as a Program Manager in 1996, trying to solve maintenance problems with the TF-34 engine. After receiving word from the Fleet that there were 61 bare firewalls, increasing at a rate of two per week, on an inventory of 117 aircraft, Massenburg called a summit of all the stakeholders - engineers, air logisticians, ICP, I-level maintenance, resource sponsors, the manufacturer's (GE) rep and others. He described the scene that ensued:

Figure 1: Enterprise Model

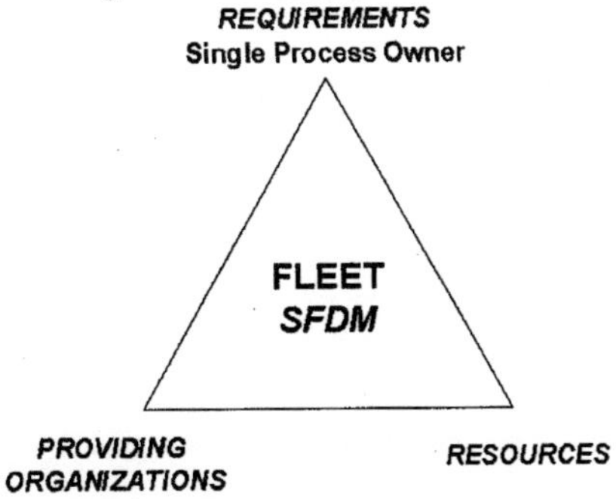

> *It was probably the first time these people had ever sat in a room together. After the second slide the place erupted, and everyone started blaming everyone else for the problems. Then somebody said: "I don't give a damn about the fleet," and all conversation stopped, because everybody at the table wanted to say the same thing. Because everyone was incentivized to look after his own territory and no one saw the bigger picture – the greater good.*

To solve the TF-34 problem Massenburg identified a leader – the single process owner (SPO) – who could be held "responsible and accountable and fireable." He then pushed the group to define the core objective – the single fleet-driven metric (SFDM) to which all parties could commit, a time-consuming process because no one was willing to subordinate their own metric to a common mission. Acceptance of the SPO and the SFDM was a tipping point, according to Massenburg, "it gets you out of the stovepipe mentality, and then the team accelerates because everyone is on the same sheet of music." Two operating practices maintained focus on the mission: communication and drumbeat. Every morning at 8:30 an email hit every computer stating the number of bare fire walls remaining. When the number reached zero, the group then focused on root cause analysis to determine the drivers of maintenance, which revealed that actual time on wing (ToW) was 500 hours for an engine rated for 1500. The SFDM was reset to "1500 hours ToW."[7]

Massenburg's triangle did not represent a formal organizational structure. It was a "behavioral model" that represented the alignment of stakeholders around a particular objective – the single fleet driven metric – and it described their behavior in collaborating to resolve process issues that cut across multiple commands. The construct enabled stakeholders to pursue a holistic "enterprise" approach to achieve a solution that optimized an objective greater than that visible to

[7] By 2006 the TF-34 had become one of the most cost-effective engines in the naval aviation inventory, with ToW at nearly 800 and rising and without a bare firewall since 1998.

individual commands. The behavioral model did not interfere with chain of command authority or responsibilities, but for the enterprise model to function effectively, participants had to understand and accept the SFDM as the overriding objective, and they had to defer to the Single Process Owner (SPO) as the keeper of the SFDM and the final arbiter of how best to pursue the mission. Participants, who included provider organizations (e.g., NAVAIR) and resource sponsors (OPNAV N88), were led by the SPO in making decisions within the triangle, and they then executed those decisions through their command structures.

Initial Success: Improved Readiness

As the SPO for NAVRIIP, Nathman initially struggled against the stovepipe mindset that encouraged commands to focus on optimizing their own set of metrics and blinded them from seeing the process holistically. One senior leader described the challenge Nathman faced in launching the enterprise approach:

> *There was an unwillingness on the part of a number of people to be subordinate in process to Nathman and take the output of their organization and line it up behind the SFDM and measure their performance in delivering against that.*

In spite of that challenge, Nathman pushed hard to inculcate the core tenets of NAVRIIP. While he could not force other commands to speed their adoption of the necessary behavioral attitudes, he did have the ability to compel change within his own organization. One critical element was changing the relationship between the fleet – the operators – and the maintenance providers. Under the NAVRIIP type model series (T/M/S) structure, wing commodores became the SPOs for their aircraft types and had to take on responsibility for readiness. Several observers pointed to the firing of a wing commodore who failed to get that message as a seminal event that encouraged buy-in to the enterprise mindset.

As NAVRIIP steadily made progress improving IDTC readiness in 2002 and 2003, leadership changes in key positions boosted the momentum and strengthened the stability of the initiative. In July 2001, RADM McCarthy took command of the Naval Supply Systems Command. In August 2002 VADM Malone relieved Nathman, who became the Deputy Chief of Naval Operations for Requirements and Programs (N6/N7), and then RADM James "Zorro" Zortman took over AIRLANT, serving as Malone's deputy SPO. Massenburg moved up to lead NAVAIR's depots in early 2003 and then took command of NAVAIR in December 2003. Zortman described the relationship between himself, Malone and Massenburg, who were all intensively involved in the formative years of NAVRIIP and the Naval Aviation Enterprise:

> *The three of us all went along together on this journey. We should not underestimate the confluence of personalities and the ability to get along and the willingness to throw the egos and ownership out the window in order to drive toward the result that we were trying to get at. There was none of this: "stay out of my swimming pool."*

Massenburg demonstrated the essence of the enterprise attitude when he took command of NAVAIR. By several accounts, his first action was to set the tone by calling Malone and stating that he was "reporting for duty." Malone professed disbelief at this, as Massenburg recalled:

> *Malone said "I don't know what the hell you are talking about. You just made 3-Star and you and I are going to run this together." I said, "No, there can't be two bosses in the enterprise; there can only be one, and I am not it. You are boss, and by the way here is $26.2 billion and 30k people and I will help you spend it." That may have been symbolic, but it was the way I thought about my job.*

Under Massenburg's leadership, process improvement initiatives at the depots coalesced and evolved into Depot AIRSpeed and then Enterprise AIRSpeed in 2003. Massenburg fundamentally changed the way the depot leadership interacted, replacing their inclination to function as independent silos, bearing proprietary attitudes about their improvement programs, with a depot enterprise approach that featured sharing of information, tools and budgets. He consolidated the process improvement initiatives – Lean Six Sigma, and Theory of Constraints - into Depot AIRSpeed and facilitated their cross pollination. In 2003 Massenburg extended the AIRSpeed philosophy and programs beyond the depots, aligning all levels of supply replenishment and repair processes – Organizational, Intermediate, and Depot - to the demands of the Fleet – creating Enterprise AIRSpeed. NAVRIIP adopted the AIRSpeed program and its process improvement tools as the enabler of readiness throughout the naval aviation community. The success of AIRSpeed and NAVRIIP in improving readiness was demonstrated by the contrasting capabilities of naval aviation between OEF and OIF (March 20, 2003). In October 2001, lack of readiness limited deployment to four carriers; two years later the Navy deployed to fight OIF with eight carriers and eight air wings (seven deployed and one ready to go.)

Adding Cost to the Readiness Equation

By mid-2003, NAVRIIP had enabled naval aviation to achieve significant gains in readiness and had connected the three levels of maintenance to fleet-driven entitlements. Those gains, however, had been facilitated in part by substantial increases in funding to support OEF and OIF. As one flag remarked, "they backed the money truck up and dumped it into the system." Naval aviation leaders began to recognize that they had achieved increased readiness but had not paid attention to cost. The impact was evident in the Navy's inability to achieve its recapitalization targets, as one leader described:

> *We were just swinging the pendulum the other way. In the late 90s we had been recap at all costs, and now we were focusing everything on current readiness. We realized that we couldn't do this without considering our impact on the future. We needed to bring cost into the equation and balance current and future readiness.*

In mid-2003, NAVRIIP adopted a new SFDM: *Aircraft Ready for Tasking at Reduced Cost.* As NAVRIIP teams uncovered the root causes of high cost, one pervasive driver was found to be that lack of reliability of components and systems encouraged operators to demand multiple quantities to ensure operational readiness. As Zortman described:

> *We didn't value reliability, so we valued inventory. We had a lot of stuff, and we didn't care how long it lasted as long as there were plenty to replace it with. And all that stuff had to be shipped, stored, repaired and modernized. So our first journey was reliability.*

One example that captured naval aviation's success in improving reliability and driving out cost started at an F18 engine repair facility at Lemoore Naval Air Station. Working two shifts and weekends, the facility was producing eight engines a month in 2002, a legacy metric based on unit output that did not consider quality or reliability. Without proper tools, sufficient parts or uniform processes, workers often cannibalized other engines in a site that was "a pigpen mess," according to one senior flag. Under the initiative of a commander with an interest in productivity tools, the facility adopted LSS and ToC techniques and within a year was producing 40 engines a month with one shift using a set of repeatable processes that led to increased quality and higher reliability. In 2004, the F18 maintenance practices were exported to the other nine F18 engine repair facilities, which were soon producing enough engines to permit the reduction in F18 sites from nine to five and eventually to three. Fewer sites with greater expertise enabled increasing reliability and longer time on wing – which in turn allowed naval aviation to reduce the total number of engines in the supply chain.

Formalizing and Expanding the Naval Aviation Enterprise

By mid-2004 NAVRIIP was making significant progress on both readiness and cost. In July, at a naval aviation leadership offsite at Thomas Group HQ in Dallas, leadership discussed possible expansion of the initiative. At the time Zortman was set to relieve Malone and RADM Denby Starling was set to take command at AIRLANT. Starling, who had been exposed to AIRSPEED training as battle group commander on the George Washington, recalled the substance of the meeting:

> *We made the decision to take the next step and start referring to ourselves as an enterprise and to start expanding the area of things we were looking at – to go beyond unit level and depot level process improvement and to try to understand what the totality of the naval aviation enterprise would be. We started talking about $40 billion of stuff that we had some influence over and how we would start looking at ourselves differently.*

Starling and several other flags wrote the one-page Naval Aviation Enterprise (NAE) charter, and, after considerable debate on word order, produced the NAE motto: ***deliver the right force, with the right readiness, at the right cost, at the right time – today, and in the future***. Structurally, NAE was identical to NAVRIIP, assuming its Board of Directors and cross-functional teams, which were titled: NAVRIIP/Readiness, Total Force Readiness, and Cost Management (Exhibit 1). Zortman was promoted to CNAF in August 2004 and assumed the role of Enterprise CEO and SPO, with Massenburg as COO. A host of other stakeholders were included in the NAE behavioral model triad (*Exhibit 1*).

Between 2004 and mid-2006 NAE's SFDM evolved from *Aircraft Ready for Tasking at Reduced Cost* to *Aircraft **and Carriers*** in 2005 and then to ***Aviation Units*** in July 2006. The latest iteration expanded the domain beyond equipment to include personnel. The critical financial dimension of the SFDM, "at reduced cost," which was absent from the initial 2002 version, was driven by the commitment of the CNAF as the SPO. Putting the people responsible for consumption (i.e. the fleet) in charge of the operation forced them to understand the factors driving cost at the various levels across the enterprise. By 2005 NAE was starting to reap tangible financial gains. That year it executed the Flying Hour Program plus Operations Enduring Freedom and Iraqi Freedom more efficiently, generating $163 million in savings that was available for other OMN purchases or for recapitalization – enough to buy two F/A-18E/F Super Hornets or 167 Tomahawk missiles.

By 2006 the NAE represented a $40.4 billion operation with more than 182 thousand people, 3,827 aircraft, 12 aircraft carriers and all associated equipment and supplies. Although the enterprise framework remained a behavioral model that was still dependent, according to several observers, on the passion and commitment of its senior leaders, supporting elements for that model were starting to be institutionalized in practices, policy and structure. For some officers, enterprise tasks had migrated from being additive work to being an integral part of their day jobs – their command duties. Several flags indicated that demonstration of enterprise behavior was becoming an important aspect of fitness reports and promotion criteria. The criteria for the Battle E award had been changed so that it was no longer consumption based. There were even several cases of changes in organizational structure to align chain of command with processes governed by the enterprise SPO. For example, Naval Air Forces took over the pilot training programs formerly managed by NETC.

Extending Enterprise Behavior Across the Navy

As NAVRIIP and NAE geared up, members of the aviation community became vigorous adherents to the enterprise approach. Many Navy leaders outside aviation, however, were either unaware or skeptical of aviation's new model. One admiral described the general reaction to seeing Massenburg and Zortman lead a "staged" walk through of an NAE Board of Director's meeting at an All Flag Officer Training (AFOTS) in 2004.

It went over my head. We were all thinking: "here are the aviators putting themselves out in front of us again. They aren't doing any better than we are; they are just doing this smoke and mirrors thing so their budget won't get cut". But that was shot across the bow, and Clark said that he wanted everyone to start behaving this way.

The AFOTS NAE demonstration was one of many vehicles that CNO Clark employed in his continuing efforts to transform naval leaders' attitudes about their "business" responsibilities and to push the Sea Enterprise initiative. He started a routine of Echelon II reviews shortly after becoming CNO, meeting with commanders to introduce a dialogue and communicate his expectations around business issues. It was apparent that, while senior leaders excelled at understanding operational issues, few had a well-developed sense of their command's business metrics or language: resource levels, costs, lines of business – working with such concepts had never been considered a command responsibility. "The goal," according to one OPNAV staffer, "was to get people to start to think about their multi-billion dollar commands as a business, and to think of themselves as business executives in addition to warfighters." One early outgrowth of the needs uncovered through these reviews was an executive education initiative launched in 2003 under the auspices of the Naval Postgraduate School.

Another mechanism to generate enterprise-like activity was the holistic review of the Sea Enterprise effort taken by VADM McCarthy after being assigned as OPNAV N4. This review restructured the focus of the Sea Enterprise effort around three pillars: Corporate level initiatives identified and coordinated through a Corporate Business Council focused on efficiencies in activities that cut horizontally across domains; cultural change fostered by expanded business education and communications; and command-level business improvement facilitated through the issuance of the Navy Performance Excellence Guidebook (NPEG). As McCarthy saw it, Sea Enterprise was the framework within which ADM Clark's vision of improved business practices would be achieved. These improved practices must permeate all levels of the Navy. In that context, he saw the emergent naval aviation behavioral model as a best practice for inculcating business enterprise practices and culture, and sought to extend its adoption across the Navy, starting with the two other major warfare domains: surface and submarine, who in their own way, had already begun to evidence some elements of enterprise thinking.

Surface Warfare Enterprise - SWE

The surface force did not experience the readiness crisis – the burning platforms of bare firewalls and tin-foiled aircraft – that plagued aviation, but it did develop a sense of urgency from its inability to answer CNO Clark's queries about its maintenance requirements and budget. In late 2002, VADM Tim LaFleur, COMNAVSURFOR, taking a cue from the naval aviation experience, hired the Thomas Group and brought together flag officers from the fleet, provider organizations, and Pentagon resource sponsors to reduce the costs associated with ship maintenance. His directive, in the words of one admiral was: "I love what you do but you cost too much money and you take the ships off line for too long. Go figure out – soup to nuts – how to do this differently." The initiative, dubbed SHIPMAIN, included maintenance, overhaul and modernization, and represented a total budget of $1.8 billion in FY 2003 for more than 160 surface vessels (aircraft carriers were added in 2004). Representatives from all the commands started to work in cross functional teams mapping out the processes – from the generation of requirements to the execution of maintaining and modernizing surface ships. VADM Sullivan, who was in charge of the modernization team stated: "We did not call it an enterprise but we were doing enterprise behavior – same stakeholders, same kinds of activity, same triangle."

Then RADM Terry Etnyre, Commander, Naval Surface Force, U.S. Atlantic Fleet, who initially served as the leader of the SHIPMAIN Requirements cross-functional team (CFT), recalled the intensive effort required to launch SHIPMAIN – myriad meetings and substantial Thomas Group ***PVM*** work. Like many other leaders, he was initially skeptical and likened the program to the

TQM/TQL initiatives of the early 1990s, which had been introduced with great fanfare but achieved modest results and soon withered. As Etnyre got more involved in SHIPMAIN, however, he saw the benefits of having a cohesive group of diverse people who were able to work together across stovepipe boundaries to align around common goals. Etnyre claimed that what tipped him over to Enterprise zealotry was witnessing the impact that process mapping and information transparency had on enabling leaders outside his command to understand the importance of their activities in a larger process and expanding their sense of ownership for their role in producing the desire output.

In early 2005, Etnyre, who had been promoted to Commander, Naval Surface Force, used an enterprise approach to oversee the introduction of the Littoral Combat Ship (LCS). The LCS was the Navy's newest class ship with significantly different manning, training and operational requirements that did not fit well into existing maintenance and training systems.[8] Recognizing that stovepiped commands were not aligned around the LCS requirements, Etnyre stood up an LCS oversight board with two cross-functional teams[9] to coordinate the stakeholders. The integrated effort developed new processes for all aspects of launching this significantly different new class of ships.

With the LCS process underway, in mid-2005 Etnyre tasked a four flag team to expand the embryonic enterprise activities of the surface forces into a full-fledged initiative. In November 2005 he stood up the Surface Warfare Enterprise (SWE) as a formal behavioral framework with three CFTS and "***Warships Ready for Tasking (over Cost)***" as the SFDM. Etnyre articulated a succinct definition of what the enterprise model was intended to do:

> *Enterprise behavior is about understanding your processes and aligning your organization to execute those processes in a way that you can monitor using metrics.*

In the first half of 2006, the SWE laid the groundwork for its operations: collecting baseline data and developing a strategic plan with defined objectives and desired effects. In the second half of the year, the SWE established the CLASSRON concept as an organizational entity to focus efforts across the enterprise and serve as an interface between the SWE and the fleet. In addition to the continued success of SHIPMAIN and other initiatives of the Sustainment & Modernization (S&M) CFT, by late 2006 the SWE's Personnel Readiness Team (PRT) had also made significant progress in aligning the training establishment to fleet requirements. A major effort was underway to transform the standard 16 week legacy training program based on a cyclical readiness model into a continuous training approach to support the Fleet Response Plan. The Phase I goal of 10 weeks and Phase II goal of 2-4 weeks promised projected increases of Warships Ready for Tasking by 75% over the FRP cycle. By late 2006 SWE enterprise activities had a projected savings and cost avoidance totaling nearly $1.4 billion (*Exhibit 2*).

Undersea Warfare Enterprise - USE

Similar to the Surface navy, submarine forces did not have a burning platform associated with readiness and cost concerns, and in several fundamental respects it differed markedly from the other warfare enterprises. Absolute priorities for nuclear-powered subs were safety and effectiveness (e.g., stealth) – at any cost. Drivers of design, operation and maintenance improvements included the loss of the Thresher in 1963, the Scorpion in 1968, [10] and collisions

[8] There were actually two different LCS models, designed by two separate corporate groups to meet stated capability and performance standards, and they featured two sets of systems that were different from each other and largely different from other Navy ships. The LCS featured three operational modules, each requiring different skills sets. The lean forty-person crew had virtually no excess manpower – every sailor was required to operate the ship almost 24/7, and all training had to be conducted ashore.

[9] Maintenance & Logistics and Manning & Training

[10] The USS Thresher, commissioned in August 1961 went done with all hands (129) April 10, 1963 while conducting sea trials 200 miles off the coast of Cape Cod. The USS Scorpion sank in the mid-Atlantic on May 22, 1968, six months after it

with Russian subs in the 1990s. Another seminal event in the mid-1990s was the loss in margin of performance superiority[11] compared to non-US subs, a condition that triggered the development of a new type of sonar under the Acoustic Rapid Cost Insertion (ARCI) program, introduced in 1998. The operational and maintenance cost profile of submarine forces was also substantially different than the other warfighting forces, especially naval aviation. A higher proportion of costs, particularly nuclear fuel, were fixed, not variable like the FHP. Maintenance tended to consist of about a dozen major projects per year, performed across six ships yards (four public and 2 private) and employing three financial systems.[12].

By virtue of its smaller size (7% of Navy forces) US submarine forces had always operated as a tighter tribe than its surface and aviation cousins. Following the 1994 OPNAV reorganization that eliminated the "Barons," who had been a locus of power (?) for each of the warfare communities,[13] submarine community stakeholders formed a core leadership team to oversee their domain. As the lead TYCOM[14], COMSUBLANT served as the head of a five person Flag Panel that included the resource sponsor (OPNAV N77), and COMSUBPAC as the principal participants, and also the head of naval reactors, the NAVSEA undersea PEO, and the detailer.

The Flag Panel, which included the key elements of the "enterprise" triangle, met quarterly to discuss maintenance issues and personnel strategy, with particular focus on retention and bonuses, and they talked weekly on the phone. Coordination within the submarine community was facilitated by the fact that there was a single operational order (OPORD) with a single set of processes for all submarines globally, an indication of the submarine forces' cohesion. Among the efforts that required an enterprise approach during the late 1990s was the ARCI initiative. The project was a spiral development effort whose use of open architecture and commercial processors, updated at much shorter intervals than legacy systems, required significant changes in testing protocols.[15] Closed circuit detailing of submarine force members was another initiative that required extensive enterprise effort.

In 2000, the Flag Panel expanded its activities, standing up two CFTs to look at depot maintenance and class maintenance plan efficiencies. It also created a dedicated training organization under the Submarine Learning Center (SLC). A functional re-alignment and creation of the Commander Sub Force title in 2003 helped to solidify the operational unity of the enterprise. When VADM Charles Munns took that title in October 2004, his predecessor had set up an off-site session with the Thomas Group to explore "morphing our enterprise to look like NAE." Munns had been N8 at CINCPACFLT in 1998 and had seen firsthand the crisis conditions that had launched AMSR, but he was not certain that the NAE model made sense for sub forces. "My view was: we don't have flying hours, we are doing pretty well financially, we have an enterprise culture and organization already. We were pretty productive."

Munns stood up the UnderSea Enterprise (USE) in early 2005, adding other Providers and Resource Sponsors to the core Flag Panel to create a Board of Directors. In an effort not to let enterprise activities become too diffuse, Munns used the BoD to prioritize projects based on available resources and projected ROI. The principal focus of USE in its start-up phase was efficient execution of maintenance projects to hit budgets and schedule targets and thus avoid the extra costs that delays incur. While much of that work necessarily involved NAVSEA, it also

---

received the briefest and cheapest nuclear-powered submarine overhaul in Navy history. (US Navy Submarine Centennial Web site)

[11] As defined by the decibel (DB) energy level. This loss may have been a contributing factor in the Russian collision incidents.

[12] Working capital, mission funded and Ship Construction Navy (SCN) (the funding stream and set of processes used for SSGN conversion and some refuelings) for new construction.

[13] In a sense the barons (OP02, 03 and 04) had been one-man enterprises, holding influence and authority over resource sponsors, providers and type commanders.

[14] The submarine forces had instituted the lead (3-star) follow (2-star) TYCOM model in 1996.

[15] ARCI enabled a 10-fold gain at 1/10 the cost.

required engagement with NAE in order to synchronize carrier and submarine repair schedules.[16] USE implemented a lead shipyard approach to identify, refine and then disseminated best practices, accompanied by life cycle management initiatives. As that objective is achieved and processes gain stability, the focus will turn toward making maintenance processes leaner and faster. A financial USE CFT was in the process of developing cost metrics on maintenance activities to support that goal.

Institutionalizing the Enterprise Framework

When Admiral Mike Mullen relieved Clark as CNO in June 2005, he came into office sharing his predecessor's enthusiasm for and commitment to transforming the Navy's business culture. Among his initial tasking orders were instructions to VADM Massenburg (the lead SYSCOM) to explore ways to rationalize the SYSCOM structure, and to VADM McCarthy to expand the activities and reach of Sea Enterprise. McCarthy and Massenburg decided to combine their taskers into a single effort, supported by Mullen, aimed at integrating the various independent activities and creating a vision for the Navy's enterprise future. Out of that collaboration came the formalization of the Navy enterprise framework, depicted in the radiator matrix and the interconnected stack of enterprise triangles (Exhibit 3 and 4). In 2006 two additional warfare enterprises – Naval Network Warfare/FORCEnet Enterprise (NNFE) and Navy Expeditionary Combat Command (NECC) – joined the three core platform enterprises and began to establish objectives and determine baseline data. Two top-level enterprises – Fleet Readiness Enterprise (FRE) and corporate Navy Enterprise (NE) -were also established and were gearing up activities.

Fleet Readiness Enterprise

The Force Commanders leading the five warfare enterprises represented the providers within Fleet Readiness Enterprise (FRE), while Commander U.S. Fleet Forces Command (CFFC), a post head by Admiral Nathman beginning in early 2005, served as the CEO. The FRE aggregated the output of the warfare enterprises – "units ready for tasking" – into "forces ready for tasking" as its SFDM. Its principal focus was delivering current readiness at best cost to the Naval Component Commanders. That objective started with an enhanced understanding of how activities impacted readiness indicators. Opportunities for cost savings could come from a variety of sources, such as improved maintenance techniques, more effective and alternative (e.g., simulator) training methods, or elimination of non-value-add tasks.

A core mission of the FRE was to facilitate the maturation process of the five warfare enterprises by accelerating the learning cycles, encouraging the sharing of best practices, and removing barriers that constrained productivity improvement. FRE focused considerable attention on pushing the enterprises to develop common process improvement methods for major activities, such as ship maintenance, which cut across USE, SWE and NAE, and involved the same Providers. One tasker in early 2007 required the enterprises to determine which maintenance practices they considered the most worthwhile to develop in common in order to drive efficiencies and speed process improvements. Similarly, the enterprises were directed to develop common performance agreements with MPTE (Manpower, Training and Education) around requirements and training to ensure optimization across the Fleet. Nathman believed that FRE's role was to steer enterprises to focus on those areas where they could have the greatest cost and readiness impacts, but he did not want to be overly prescriptive.

> *People have to live this experience – sitting around their own triangle enough so they start seeing for themselves what needs to be done. We need to empower and enable the Force Commanders. If this becomes too top-down, we will kill it. If the Force*

[16] Carriers generally (though not always) tended to receive priority and overlapping shipyard stays of a carrier and a sub could cause major delays for one or both.

Commanders don't feel like they own it, we will lose the goodness of the behavioral model.

One key to empowering the enterprises was creating effective incentives, and FRE instituted several measures around financial systems that increased the Force Commanders authority and encouraged them to identify and capture savings. During FY 2007 the FRE began the process of assigning responsibility to individual force commanders for execution year decisions about the allocation of resource dollars, to be institutionalized with formal transfer of 1517 authority in FY 2008. FRE would maintain visibility and retain approval authority for exceeding spending thresholds or changing requirements, but essentially the checkbook and financial operations would be decentralized and handed down to the commands to execute. Nathman's vision was to amplify the incentive of the 1517 authority transfer through retention of execution year savings achieved by the enterprises and allow Force Commanders to use those savings to spend on their own enterprise priorities.[17] However, models used to determine programming requirements would be updated to reflect the efficiencies realized by the enterprises, aligning future year resources to desired readiness output. To support these changes, senior leaders were being provided opportunities to improve their financial and business skills, such as through the Navy's Executive Business Course and the creation of financial workshops. Mark Honecker, FFC's Executive Director, spoke of the change:

This represents the removal of a significant barrier. The existing system is a disincentive; a chief complaint from commanders is that we had not aligned authority with responsibility. This transfer of authority empowers them to develop a better understanding of where they spend their money, and now they will get to keep the money they save in execution and reinvest it to achieve greater efficiencies.

Navy Enterprise

The missions of the broader Navy Enterprise (NE) were to set business strategy and policy, provide the governance forum to deal with both current readiness and future capability issues, and to act as the Navy's top-level barrier removal board when senior level arbitration is required. The NE framework was adopted by CNO in mid-2006 and was activated in late 2006. RDML David Buss, who had been engaged in various facets of NAE since 2003, was selected as Navy Enterprise Chief of Staff to assist CNO in organizing and initializing the activities of the Navy Enterprise Executive Committee (NE ExCom). Buss's initial focus was on facilitating the establishment of the VCNO, ADM Bob Willard, and the Assistant Secretary of the Navy for Research, Development and Acquisition, Dr. Deloris Etter, in their roles as leaders of the Provider Enterprise portion of the broader Navy Enterprise framework. This necessitated a series of Echelon II visits to each of the nine Provider commands, conducted from August 2006 through January 2007, with the purpose of baselining each Provider: mapping their resources (people, dollars, and inventory) and processes in support of their customers, the five warfare enterprises. The meetings underscored the critical importance of terminology and the on-going need to establish a common enterprise lexicon. Another element of those early efforts was establishment of a performance agreement framework for each warfare enterprise-Provider intersection in the Navy Enterprise matrix in order to establish accountability and set expectations of what was needed to generate the requisite level of readiness output. Other initial activities included identification of "corporate" metrics and development of a strategic communications plan, whose early products included NE's first NAVADMIN, which formalized the creation of the warfare enterprises, as well as a CHINFO "Rhumb Lines" article and several CNO communiqués to senior Navy leadership.

[17] The calculation of savings was complicated by the fact that some initial level of savings was required just to meet cuts ("efficiency gains") already incorporated into programs. Further savings would effectively be retained by an enterprise for two years, after which changes from the realized efficiencies were reflected in programming models.

One of the Buss's early challenges was establishing an NE governance model with an appropriate drumbeat for senior leadership. Given the commitments and schedules of the Navy Enterprise's senior leadership, which included CNO, VCNO, CUSFFC, and several Assistant Secretaries of the Navy, Buss's goal was to incorporate enterprise-related discussions and activities into a new Executive Committee forum, which would act as the Navy's "corporate board." The new model would enable the eventual discontinuation of several legacy forums and processes in order to avoid consuming additional bandwidth of senior leaders.

In mid-November 2006, the NE ExCom held a dedicated two-day retreat to coalesce senior leadership thinking around the definition and implementation of its mission: articulating a vision for how a mature enterprise would act and determining objectives, priorities and desired effects. The group spent considerable time discussing the relationships between the various enterprise levels and defining "lanes" for NE, FRE, Warfare Enterprises, and Enabling Domains. Since February 2007 the ExCom has maintained a drumbeat of meetings whose tempo is dictated in part by subject matter but which occur on a near bi-weekly basis. One regular agenda item is a CNO Monthly Review (CMR): a detailed look at current readiness and future capability programs designed to give CNO an "OPNAV view" of issues prior to the SECNAV's Monthly Review (SMR). The ExCom is also beginning to deal with strategic "corporate" issues such as NECC Wholeness, Next Generation NMCI, and programmatic issues associated with the PR09 Budget.

Among the key challenges that the NE faced were figuring out how to apply the enterprise model to acquisition programs and what the corresponding enterprise relationship between the Secretariat and uniform leadership would look like. Nathman expressed a strong opinion about keeping a clear distinction between current readiness (CY and CY +1) and future readiness related to acquisition strategy and funding.

Exhibit 1: Naval Aviation Enterprise Organization

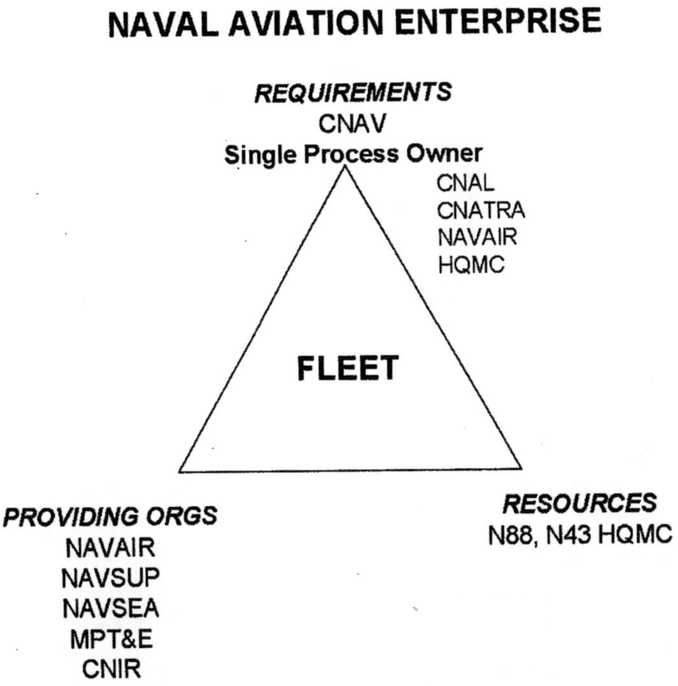

Exhibit 2: Total Projected SWE Savings and Cost Avoidance (as of October 2006)

Initiative	Projected Savings	Projected Avoidance
DDG Cost Reduction		$267.1M
Geographic Detailing		$2M
eRMS		$2.0 – $2.5M
NEC Mismatch Reduction		$10M
Diesel Improvement Program	$50M/year	
SHIPMAIN	$690.4M	
Customer Relations Mgmt	$25M	
METCal Mgmt System	$10M	
Distance Support/Tech Assist	$5M	
Performance Based Logistics	$300M	
FY06 CNSL Fuel Under burn	$13M	
FDNF NEC TADTAR	$.25M	
Total	**$1,093.65M**	**$286.1M**

Exhibit 3: Navy Enterprise Framework

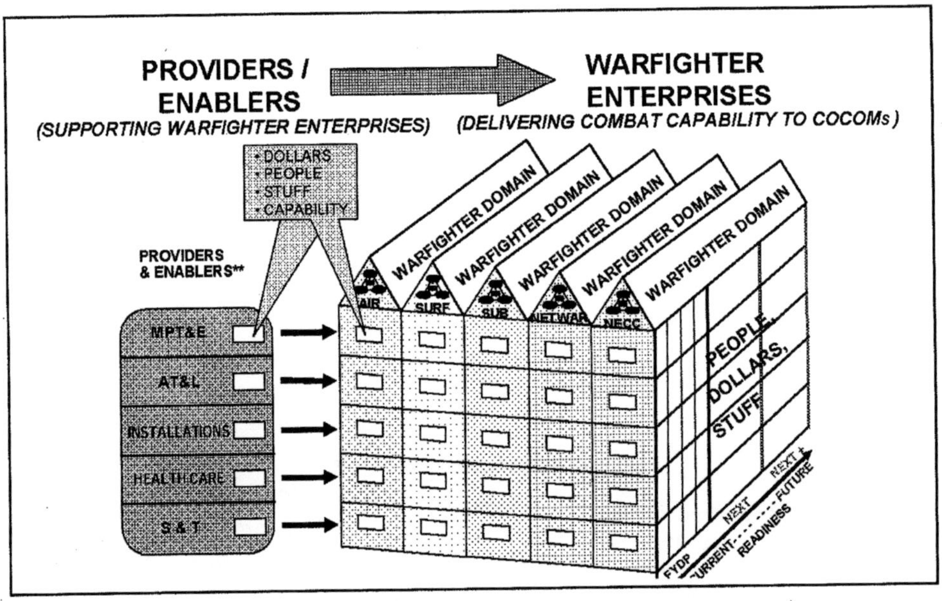

Exhibit 4: Navy Enterprise Framework

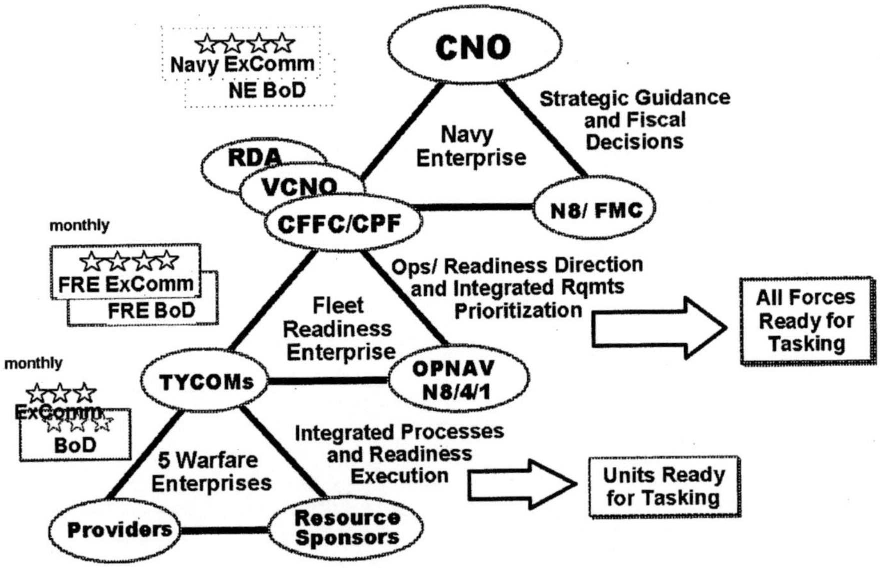

Appendix 1: Process Improvement Tool Definitions

Lean is a process improvement strategy that focuses on the ability to make everything, everyday in the exact quantity required, with no defects. The goal is to achieve perfection through the total elimination of waste in the value stream of the process. Lean uses incremental improvement to constantly expose waste to balance operational and standard workflows. Most notable examples are the supply chains established by Toyota and Honda.

Six Sigma is a process improvement strategy that is based on the assumption that the outcome of the entire process will be improved by reducing the variation of multiple elements. Six Sigma is uniquely driven by a close understanding of customer needs, a disciplined use of facts, data, statistical analysis, and diligent attention to managing, improving, and reinventing business processes. Six Sigma focuses on variation reduction to produce highly repeatable processes that create customer satisfaction. Six Sigma is a measure of variability in relation to a total population of numbers.

Kaizen is a method for accelerating the pace of process improvement in any setting. It is typically a rapid (one week or less) intense event where progress is made through all of the Define-Measure-Analyze-Improve-Control steps. This requires preparatory work completed on the Define step and even sometimes the Measure step by a small team led by a team leader and a Black Belt. The rest of the work is done by a full group where the participants work ONLY on the project and are relieved of responsibilities.

Theory of Constraints (TOC) is the process improvement and systems thinking skill based on the belief that any organization has at least one constraint and that any improvements on non-constraints may not yield as significant return on investment as working on the constraint.

Process Value Management™ (PVM™) is a 'tops down,' process-based, cross-functional change management methodology based on the identification and systematic removal of business process and cultural barriers. Improvement results are process metrics-driven through a top management leadership team Board of Directors. ***PVM*** is based upon Thomas Group's proprietary private sector change management methodology that has been successfully imported into DOD readiness production processes. As multiple processes in large organizations are improved, they are linked and aligned with advanced ***PVM*** techniques to produce public sector enterprises like NAE and SWE.